THE
GRID

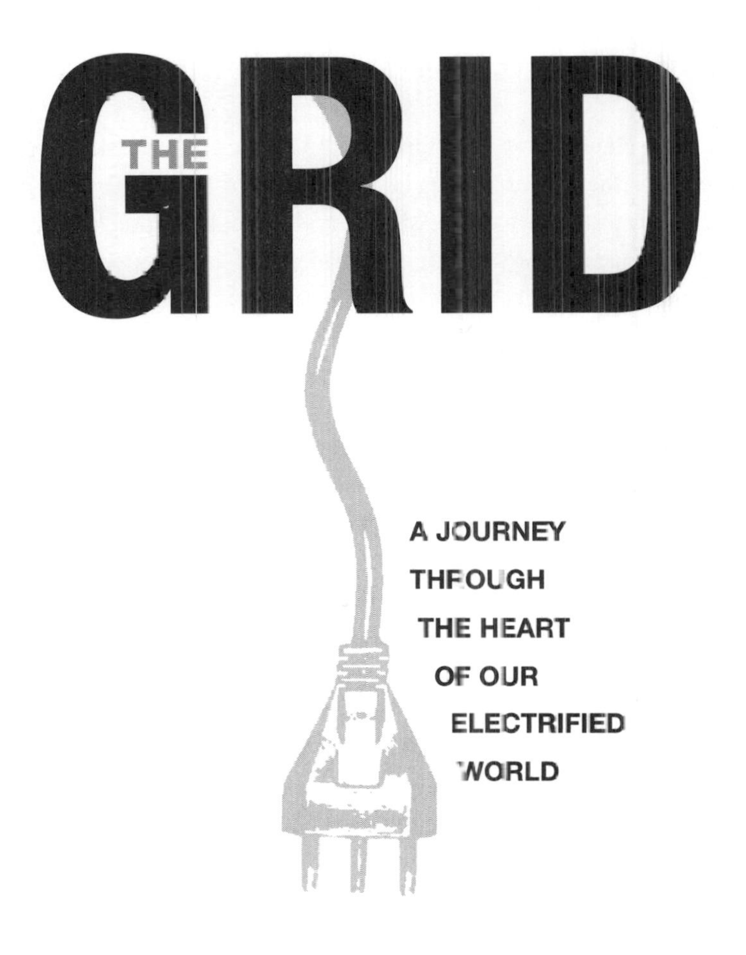

PHILLIP F. SCHEWE

JOSEPH HENRY PRESS
Washington, D.C.

Joseph Henry Press • 500 Fifth Street, NW • Washington, DC 20001

The Joseph Henry Press, an imprint of the National Academies Press, was created with the goal of making books on science, technology, and health more widely available to professionals and the public. Joseph Henry was one of the founders of the National Academy of Sciences and a leader in early American science.

Library of Congress Cataloging-in-Publication Data

Schewe, Phillip F.
The grid : a journey through the heart of our electrified world by / Phillip F. Schewe.
p. cm.
Includes bibliographical references and index.
ISBN-13: 978-0-309-10260-5 (cloth with jacket)
ISBN-10: 0-309-10260-X (cloth with jacket)
ISBN-13: 978-0-309-66279-6 (pdfs)
ISBN-10: 0-309-66279-6 (pdfs)
1. Interconnected electric utility systems—History. 2. Electric power distribution—History. 3. Electric power production—History. I. Title.
TK447.S34 2007
333.793′2—dc22

2006029231

Cover design by Van Nguyen

Printed in the United States of America.

CONTENTS

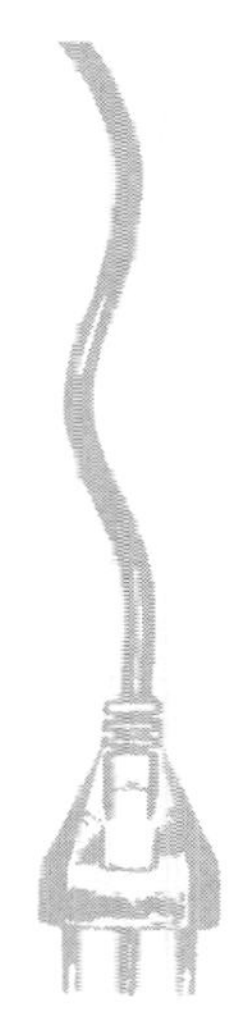

INTRODUCTION

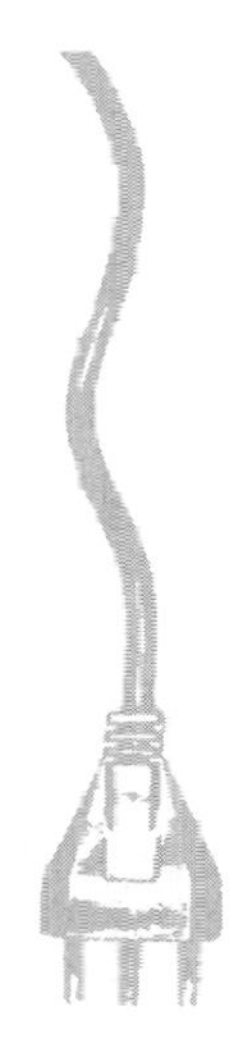

The electrical grid goes practically everywhere. It reaches into your home, your bedroom, and climbs right up into the lamp next to your pillow. It's there while you sleep, and it's waiting for you in the morning. Taken in its entirety, the grid is a machine, the most complex machine ever made. The National Academy of Engineering called it the greatest engineering achievement of the 20th century. It represents the largest industrial investment in history.

And yet the electrical grid is not a single *thing* but several things: a highway for delivering a product to millions of customers, a sort of NATO defense alliance of utilities pledged to help each other in time of need, a platform supporting a worldwide movement of information, and a commodities exchange dispatching vast resources on a second's notice. The grid almost seems alive, like some enormous nervous system.

Electricity is, in essence, a form of bottled lightning. *The Grid* is partly about the bottle—restless energy contained, sometimes only barely, in a colossal network of wires and machines—and partly about the lightning—a domesticated version of the thunderbolt that roams the sky, a tamed potency metamorphosing here and there as heat, light, torque, traction, and bits in a digital stream.

This is the first full-length popular book devoted to the development of the power supply system from Edison's era right up to the present day. It is history but not academic history. It does not strive for comprehensive coverage of technological milestones. It does not recount the history of the grid uniformly on all the continents but rather concentrates on power development in North America. The emphasis is on telling in simple and vivid terms the *story* of electricity.

The coming of the grid was the engineering equivalent of homesteading. Hardy pioneers pushed out into the unknown prairie of technology, first establishing small hamlets that later developed into great metropolitan zones of energy. Later still came the universally connected energy network taken for granted nowadays in the developed world. In poorer countries—and this accounts for a majority of the world's people—electricity is scarcer. The mighty effort to enhance living standards through efficient use of energy is at an early stage, alas, even now.

Indeed, although the electrical grid has become irreplaceable and vital to many aspects of public and private life, it comes at a heavy price. One is the cost of business losses stemming from power outages. In the United States alone this amounts to tens of billions of dollars per year. There is also the political cost of acquiring energy supplies from unstable countries and the environmental cost of extracting fuel from the earth and sending pollution into the air, water, and soil.

The book starts with that great shock to the system, the costly August 2003 electrical failure in North America, an event in which 50 million people lost power. What happens to a major city like New York when elevators, trains, traffic lights, gas pumps, and heart–lung machines suddenly stop? We usually take electricity for granted and don't realize how important it is until it's gone. Chapter 1 looks at how far the tentacles of electrical convenience have insinuated themselves into modern living.

The actual onset of the grid, through the buccaneering efforts of entrepreneur-inventors such as Thomas Edison, George Westinghouse, and Nikola Tesla, begins in Chapter 2. The laying of the keel for what would later become the worldwide grid began in lower Manhattan as an act of discovery and intrepid exploration that could be compared in spirit and importance to Columbus's arrival in the New World. Along the way you will learn what electricity actually is. You're in for a surprise.

Chapter 3 shows how electricity flowed for the first time into the homes of ordinary people, not just the rich customers served by Edison's early grid. And when electricity arrived, it produced not one revolution but several—in domestic arrangements, in the way factories are run, and in the way information and products are designed and delivered. The birth of electronics ushered in the age of radio and then television. As a result, evening entertainment changed a lot.

Chapter 4 shows what happened when electricity finally vanquished its energy rivals. We look at the development of three large electrical networks: the grid in the Soviet Union, born amid revolution and civil war; the grid in the Tennessee Valley, developed during the Depression and World War II; and the grid in New York City, where Consolidated Edison became the most famous utility in America.

Chapters 5 and 6 look at what I will argue is the worst day in grid history, November 9, 1965, when a massive power failure in the United States and Canada—an earlier version of the 2003 blackout—swept electricity away from 30 million people. Chapter 7 examines the long aftermath of that event and the extensive restructuring of the power industry, chiefly through divestiture, deregulation, and the introduction of new technology. The effects of this huge business shakeup are still unfolding.

How is electricity made? Chapter 8 visits several power-producing venues, including "Big Allis," at one time the largest steam-electric generator in the world. Then we inspect an energy-efficient home in Maryland that not only produces most of its own electricity but even sends some back into the grid. Finally, we'll tour a nuclear power plant on the Hudson River, where security is very tight.

How do you get electricity where you live? By seeking out and examining the workings of a single typical energy company, in this case Idaho Power Corporation, Chapter 9 shows what happens behind your walls and in your wires. You will see how power is generated; how it's transmitted over long distances; how the grid gets repaired; how energy gets sold, divided, and resold; how it comes into your home; how it gets into your rooms and appliances; and how corporate managers determine how electricity will be made 10 years from now.

Many people possess the benefits of electricity, but many more do not. The intricate final chapter compares power grids in several places—Ohio, Uganda, India, and China—and considers how electricity and quality of life are related. The book ends not with predictions of future energy policy or summaries of pending high-tech breakthroughs—such predictions often misfire and, anyway, the technology of today will soon be overtaken by newer developments—but by a notable example of human ingenuity at work, as manifested in electrical devices and engineering daring. This shining event, which helps illuminate the meaning of the grid, is the moon landing of 1969.

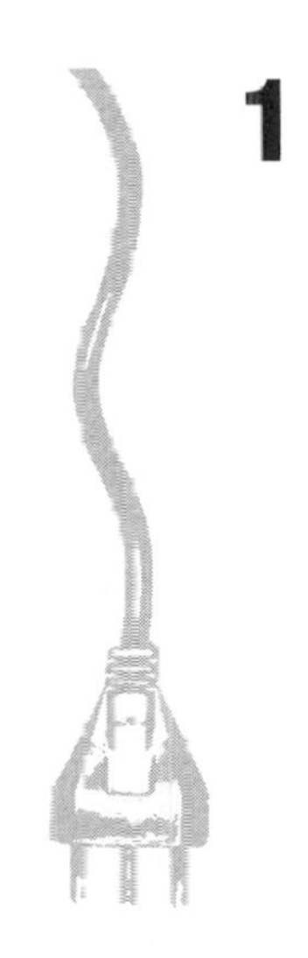

1

THE GRIDNESS OF THE GRID

The airborne view of New York City is surely a picture fit for the gods. Strapped into your seat with no escape you might be a bit nervous, but what you see out the window diverts you from apprehension. There it is on display, a habitat for 20 million souls. The largest contours you can make out by the light of day—Manhattan's fabled water frontage, the inside curve of Coney Island, the bushy green of Central Park—are due in part to purposeful digging up or filling in. Other objects in your visual field are more obviously man-made: the tall buildings, airfields, and oil refineries. The railroad grid is harder to spot; you need to look for one of those telltale fan-shaped rail yards and then trace the radiant lines outward from there.

At night the view is even more splendid because that's when the electrical grid stands out. Buried in the earth and hidden from view while the sun's up, the grid reveals itself after dark in brilliant fluorescently lit rectilinear streets and diagonal boulevards, all geometric proxies for the electric current running beneath. This luminous quilt is practically continuous around the horizon. Even at the watery margins of the metropolis—the rivers, the bay, the sound—where technology usually shuts down, the electrical grid, assisted by bridges and causeways, runs irrepressibly onward.

We linger over this overhead shot of the Big Apple since, besides being one of the most arresting sights in the world, it illustrates more directly than any view from the ground the ubiquitous physical presence of the grid. It's where the power industry in America got started and where probably more electricity gets consumed than in any other major city of the world.

WITHDRAWAL SYMPTOMS

On August 14, 2003, something happened that brought millions of New Yorkers out into the streets. They came up from underground tunnels and down from upper-story offices to see what had gone wrong. And at the same moment millions more came forth in Detroit, Toronto, and Cleveland. No, it wasn't *The War of the Worlds* or an atom bomb but something else, something electrical.

Imagine you'd just come from an office meeting, picked up some documents, stepped into an elevator, descended for some seconds, and then for a reason not apparent, everything abruptly halted, leaving you dangling in the dark in a metal cubicle the size of a luncheonette booth. Here's what you might have told yourself: Next time, I'm going to take the stairs. Why do they make buildings this high in the first place? Why did it have to be *this* elevator?

When electricity stops, one of the less fun places to be is an elevator stalled between floors. It's difficult to say which is worse: the growing heat, the skimpy ventilation, the tight space, or the lack of light. Maybe worse is the suspicion—at first just a queasy feeling, then a looming fear—that gravity could easily finish the job at any moment, overcoming whatever fragile restraint is holding the elevator in check, allowing it to rush to the bottom of the shaft with an acceleration of 32 feet per second every second. And so, instinctively, some passengers forebear to move at all, as if too sudden a gesture would tip the hidden balance and hasten the rapid ride down. No one forced me to get into the elevator, you would say to yourself. It was there and I took it.

On August 14, hundreds were stuck in New York elevators. Some waited it out quietly, while others screamed. One Wall Street lawyer found himself in an elevator in the dark next to a screamer, a woman who felt sure their predicament was related to a terrorist attack. Keep in mind that this particular summer day came just short of two years after the World Trade Center attacks of September 2001 and that for many New Yorkers in a tall structure only a few narrow blocks from Ground Zero a jarring failure of the machinery could reasonably seem to be the prelude to catastrophic building collapse.

The lawyer happens to be blind, so on this occasion, visually speaking, he is in his element. He could take control of his emotions and calm the frightened woman. Smell. There isn't any fire; there aren't any fumes. Listen. There isn't any frantic commotion. Feel. There is no crashing or shaking. We just happen to be stopped between floors. That's why everything is muffled. Try pulling back the doors and determine where we are. It's dark, but we'll make it out.

Sure enough, after some exertion they got the door open. A short drop of a few feet down into the lobby at the 32nd floor and they were free. Using his cane to tap the way and no worse off because of the dimness, the lawyer proceeded down the stairs the rest of the way with everyone else and out onto the streets filled with the questioning throngs.[1] They, and millions of others trudging down emergency stairwells, streaming along clogged thoroughfares, and teeming across bridges, would soon discover that their local commotion was part of something bigger. In fact, it was the largest power failure in history.

If you wanted to locate the cause of the 2003 blackout, you could go backward in time a few hours and a few hundred miles west to Ohio. Or you could go further back and look for imperfections in Thomas Edison's original New York grid laid in the earth more than a century before. Or if you wanted to search for the true origins of human electrical habits, you would need to harken further back. In those remote centuries the Chinese crafted their knowledge of magnetism into a workable, direction-orienting device—the compass—and yet there was no Chinese grid. At the opposite edge of Eurasia, the ancient Greeks played with magnetized rocks and with staticky tree resin (*elektron* was their name for amber, which could attract certain materials if you rubbed it with fur), but there was no Greek grid. Not until the 20th century was there grid enough and did electricity go far enough for the inconvenience brought on by an electrical failure to reach the million-person mark.

On August 14 there was much inconvenience. Because all electric trains in town refused to budge, you might, for example, have been forced to spend the night sleeping on the steps of the main New York City post office. The building's pillared front stretches for two city blocks. It's straight out of Rome but bigger. During the day thousands trudge up the steps on postal matters. At night the same marble can become the resting place of ragtag slumberers nestled on skids of cardboard. The upstanding wouldn't think to linger there.

Yet that night crowds were present, not the few regulars but hundreds with nice dwellings elsewhere. The sleepers were not vagrants. Some were svelte in heels; a few possessed law degrees; most had plentiful cash on their persons. Although they might otherwise have hesitated to surrender to sleep—not in public, not on concrete, not with their side-pocket wallet a tempting target—here they were nevertheless because no preferable alternative presented itself. Water bottles, paper cups, backpacks used for pillows, the recumbent heads of sprawled strangers—there was hardly a place to step. The post office had become a stadium of sleep.

Across the street, a honeymooning couple from Kentucky snoozed fitfully in a bed of flowers fronting Madison Square Garden. Not the ideal nuptial spot, but people sleep where they can. Ejected from her hotel, which had shut down, a teacher visiting from Florida spent the night on the sidewalk. "I felt like a street person," she told a roving newspaper reporter. So near and yet so far; a man who lived just outside the city and who had exhausted all evacuation alternatives bedded down in Bryant Park behind the main library building, the one with lions out front. There he slept with a thousand others. "I don't recommend it" is all he had to say.

New York City was the biggest but not the only victim of the 2003 blackout. Across numerous midwestern and northeastern states and in portions of lower Canada, the electrical grid had rapidly drained for reasons not then known. Technology itself had not collapsed, but the energy needed to keep things humming had disappeared. In consequence, rotating machines had slowed to a halt, heating machines cooled, cooling machines warmed, broadcast machines grew quiet, and information machines went dumb. The event was a blizzard of deficits—a deficit of energy, a deficit of light and motion, and, especially for those pondering a night's sleep around Madison Square Garden, a deficit of options.

An event like the 2003 blackout offers a rich opportunity for standing outside the electrical grid in order to assess its importance. After all, what does a fish know of water until removed from its aqueous element? That's why, like sightseers getting a better view of a plain by climbing up a mountain slope, we've begun this tour of our electrified world by first reaching a vantage point somewhat *outside* electricity, at least for a short while. This is possible only during electrical failures, when grid routine is interrupted. Most people view blackouts as highly inconvenient, bad for business, and potentially dangerous to life and limb. Nevertheless, it is precisely from within these brief but dramatic black holes that we can get a visceral feeling for the octopus overgrowth of high-voltage technology.

Do people really need to be reminded that a massively engineered quilt of energy-filled wires is on duty inside their walls? Yes. So great is the prevalence of electrical infrastructure that at times our powered devices might seem like part of the environment. The grid may as well be a geological formation. We neglect to think of electricity as a *made* thing, a costly substance deliberately delivered to a specific location. Rather it comes to take on the aspect of ambient reality, a kind of bubbling organic hotspring of versatile sustenance. For someone on the 32nd floor of a tall building, a bank of elevators might seem like no more than the obvious and best path down from the heights. It would take some effort of imagination to see the elevator and the building as an artificial trail inside an artificial mountain.

A sense of entitlement gets the better of us. How could an elevator, which had worked all those other times, not work now? Without advanced mathematics it's difficult to perceive a total electrical failure as one outcome along a spectrum of possible occurrences. It's more instinctive to feel *wronged*—being detained in an elevator is not a happenstance but an affront.

The victims inside a frozen elevator must wonder whether outsiders, in the part of the world still electrified, even know of their predicament. To tighten the screws a bit more, let's specify that you have no wireless phone, and that the elevator intercom is broken, a warranted assumption. So in addition to the lack of light, ventilation, and movement, you can add lack of information to the list of things lost when energy flows out of the grid. It can be disturbingly dark in an elevator, more profoundly dark than most people have ever experienced. One thing you can do to cheer yourself up is to strike matches in an effort to push back the darkness. But then some clever person observes that the available oxygen might be in short supply, and so the hostages voluntarily lapse back into the dark. That's when exact spatial coordinates become blurry and the flow of time gets pulled like taffy by the forces of disorientation to an extraordinary degree. If it weren't so awful it would be wonderful.

CLOSING THE CIRCUIT

When it first started to be handled a few centuries ago, electricity began as a novelty and an amusement. The early practitioners were part scientist, part vaudevillian. They wanted to learn but also wanted to show off. How better to publicize the new electrical knowledge than with a demonstration before a live audience? In 18th-century France, in the stylish salons typical of the Enlightenment, you might have seen a chain of volunteers linked hand to hand, some of them ladies in brocade gowns. Together they constituted a sufficiently conductive path to convey and illustrate the electrical phenomenon. If the built-up charge on the generating device were large enough, a mild jolt, maybe even a spark, would result, dramatically showing that electricity was not entirely a static thing. It didn't have to sit in place. It could move. Electricity consisted of the presence or movement of a mysterious substance called "charge." Charge could spread itself a mile if you had a metal wire that long. You couldn't see charge or weigh it; you could only detect its coming or going and feel its zing. It was like tasting the tartness of a lemon without having the lemon.

One of the leading savants to glide through Parisian high society was an American, Ben Franklin. Long before he came to edit Thomas

Jefferson's draft of the Declaration of Independence, Franklin edited the electrical experiences of mid-18th-century electricians, and he generally gets the credit for suggesting that there are not two types of electricity, attractive and repulsive, but only one and for instituting the placement of lightning rods on buildings to ward off the capricious ravages of atmospheric charge.

And then there was Franklin's little experiment. More dramatic than any parlor trick was nature's own way of getting rid of an overabundance of charge in the sky, some of it lingering on the undersides of clouds, through the agency of a sizzling charged stroke of lightning. The proof that lightning is a form of electricity has entered into common folklore through Franklin's famous demonstration—he was lucky not to be killed in the process—in which he induced a passing thunderstorm to discharge its excess charge down to Earth through a conducting thread carried aloft by a kite.

What we know now, and Ben Franklin did not, is that electricity involves the flow of trillions of tiny particles. More than trillions: a typical 1960 air conditioner ran on a current of about 7 amperes. This turns out to be about 40,000,000,000,000,000,000 electrons passing by every second. This is an absurd number—small in comparison to the number of charges associated with lightning and yet large in relation to the pool of charge collected in those early electrical games. The ancient Greeks, by rubbing fur on glass, could isolate only small amounts of charge. Ben Franklin and his colleagues could separate far more charge with their hand-cranked machines. To make electricity really useful, though, you can't just move some charge from one place to another. You have to *keep* it moving. To turn motors, lift elevators, illuminate bulbs, toast bread, energize trains, electricity must move as a continuous current around circuits.

For that we can thank Allessandro Volta, who completed the circuit. Around the year 1800, he was experimenting with different arrangements of metal discs in contact with a container of chemicals, when he got charges to flow, and keep flowing, through a conducting wire. Volta had invented the first battery. Small in size, weak in energy, trivial in practical effect, Volta's setup wasn't exactly a grid. Nevertheless, it counts as the first tryout for later grid development. In this book you're going to hear of far larger machines for producing flowing electricity, but Volta's battery was the first. You will read a lot about circuits extending hundreds of miles, from Ohio into Michigan, over to Ontario and back south into New York, but Volta's tabletop loop, consisting of a simple wire hooked up to the ends of his battery, was in effect the first circuit.

On August 14, 2003, the Ohio-to-Ontario circuit, a colossal magnification of Volta's loop, was in place but nearly empty. In a blackout not only the electricity stops. Cash stopped in Newark because automatic teller machines depend on power. Gasoline stopped in Ottawa because the pumps at service stations are powered by the grid. Underground mass transit in Toronto stopped. Air conditioning in Buffalo stopped, which was unfortunate since August 14 was hot. Conversely, the stopping of the grid can cause other things to start or speed up: waiting, sweating, complaining, walking, and the thawing of frozen food.

Water can stop. In Detroit electrically powered filtration stopped, so no more fresh water came through the pipes. In Cleveland the pumps that pull water out of Lake Erie stopped, so there was no more drinking water. The associated process of waste treatment also stopped. Millions of gallons of sewage poured backward into the lake. Not surprisingly, swimming in Cleveland stopped.

Cities are hungry beasts. They have factories for making things, mile-long avenues of stores and showrooms for selling products, and fat phonebooks of agencies for providing services. A majority of the raw materials needed for sustaining this gargantuan metropolitan metabolism, however, must come in from outside. Cities consume vast amounts of food, fuel, water, and finished goods that arrive day after day by plane, ship, train, truck, pipe, and wire. Much of this flow halts when the electrical grid halts.

One of today's most precious commodities, electricity hardly figured in the life of cities before the end of the 19th century. Electricity played no role in the administration of Imperial Kyoto or the meteoric rise of Islamic Cairo. Electricity came too late to inspire the artists of Renaissance Florence or contribute to the glory of Paris in the time of Napoleon. Once it arrived, however, it became a big factor—some would say too big a factor—in the rhythm of urban existence.

The advent of electric power and its catalog of follow-on products radically changed the way many things were done: the architecture of buildings (which no longer had to be configured to maximize sunlight pouring in at the windows), the sprawl of suburbs (cheap electric trolleys allowed employees to live farther from work), the methods for storing and preparing food (perishables could be maintained for days, frozen goods for months), the organization of factories (which didn't have to be situated next to a river and could be kept open for night shifts), and the lighting and heating of homes. Fairly, we could argue that much of what we call modernity is fundamentally electrical in nature or at least dependent in a fundamental way on the electrical grid.

Modernity is an empty term to the person behind the steering wheel of a car that has gone scarcely three blocks in an hour. When electricity stops, so do the traffic lights. And so does traffic. All the road conventions we willingly obey in the interest of safety and harmony—green for go, red for stop—are suspended when the power grid gets suspended. So it was on August 14 when in dozens of city centers the traffic molasses known as gridlock quickly took hold. If only someone would step in and direct traffic by hand. One person, telling other people when to halt or wait or proceed, could make all the difference. It wouldn't have to be a sworn police officer. You just step from the safety of the curb and into harm's way. The rhythm comes quickly. With one hand you order some cars to advance; with the other hand you halt the opposing column of cars. The motorists, happy to break out of their stalemate, are practically unanimous in their gratitude.

There *are* things worse than being stuck in an elevator or in traffic. You could, for instance, be undergoing surgery when the power leaves. Or you could be a patient hoping for a life-saving organ transplant when, after an anxious multimonth wait, a suitable liver arrives. The operation is about to begin, and *then* the power goes out. On August 14 there was just such a case. The liver, allotted to a patient in New York City who (in the absence of electricity) could no longer safely receive it, was driven and airlifted to Pittsburgh, where it was successfully implanted in someone else.

Some blackout experiences certainly amount to more than mere inconvenience or anxiety. When the grid stops, oxygen can stop. Most of us breathe easily, but for those with asthma, emphysema, or lung cancer, an extra flow of oxygen is sometimes prescribed. Hospitals usually have backup supplies, and attending staffers can see to patients' needs. But what about respiratory outpatients? In homes extra oxygen can be produced in machines called concentrators, which separate the part of air made of nitrogen from the part made of oxygen. What happens when the machine you need for breathing itself runs out of breath? Many patients, stranded at home, have cylinders of oxygen in reserve, but these last only a few hours. What then?

On the night of August 14 one Michigan oxygen company had hundreds of home patients on ventilators, monitors, and other machines that stopped working for want of power. Detroit and Ann Arbor were particularly hard hit, and company drivers from all over the state, some even convoyed up from Indiana, went into action making calls on those who needed extra supplies. The rescue was successful.[2] Everyone who needed oxygen got it.

In the year 10,000 BCE, or at any other time during the last 99.9 percent of human existence, automobiles, elevators, and traffic lights weren't around. To be in a dark place was the natural state of affairs, while to possess discretionary light was remarkable. Because of the electrical grid, all this is reversed. Having light all the time is the custom, and going lightless becomes a memorable anecdote. We need the grid. We count on it. When thousands of people get stalled in various metal conveyances moving horizontally through the city in subways or vertically in elevators and are made prisoner by the failure of a wire in Ohio, we are incredulous. How did this state of affairs come to be? I don't mean the particular interruption of the grid in 2003—we'll look at that later. I mean the grid itself, and not merely the metallic part of the grid alone. It won't be enough to recount the invention of electrified machines without also considering the impact of those machines on human life and on the planet.

WALDEN INCORPORATED

Consider them side by side: Michael Faraday and Henry David Thoreau. My intended dual historical perspective, technological and social, is exemplified by these two men, who flourished in the middle third of the 19th century. One was a great physicist and inventor, the other a great philosopher and poet. Faraday is famous for what he saw out of the corner of his eye one day in his cluttered laboratory in London. Thoreau is famous for the two years he spent living at the shore of a small lake not far from Boston and for the book he wrote about the experience. Faraday discovered electromagnetism, and Thoreau wrote *Walden*. Electromagnetism is the tremendously useful principle at the heart of electrical phenomena, and it drives the trillion-dollar electrical industry. *Walden* is a memoir about living away from technology. A *Walden*-related industry also exists, if I may so name the movement dedicated to evaluating or promoting life values (things like leisure time, rumination, family togetherness) even as our lives become more and more entwined with machines, but it cannot compete in dollar value with the electrical industry.

Indeed, in this book Faradayism will outweigh Thoreauism, at least in terms of pages. The discovery of scientific truths and the building up of the electrical grid as a practical energy distribution scheme will predominate over philosophical questions about the grid. But the questions will be there. This book takes the view that the electrical grid is generally a good thing. Few people who have access to electricity would willingly give it up. The story of the grid will mostly be a story of buildup, progress, enlargement, innovation, even revolution. But the grid is not an unalloyed good. It has brought problems in its wake, as the blackout illustrates. Faraday, along with the engineers and managers who built the grid, performed a mighty task, but no less important to the grid saga will be Thoreau and poets and historians who help tell us what our technology *means*.

Before we see what Faraday and Thoreau did, let's look at what the Danish scientist Hans Christian Oersted did. In 1820 he noticed that an electrical current flowing through a wire creates, or *induces*, magnetic forces in the surrounding area. How did he know? Because the current caused a nearby compass needle to swing about. This was, at a single stroke, a great scientific discovery—electricity can cause magnetism—but also a great technological development, or soon would be. If you sent enough current through a coil of wire wrapped around a piece of iron, you could create a powerful magnet. With an electromagnet you could lift hundreds of pounds, even thousands. In ways like this, puny humans gain extra leverage over their environment. And that's often what we want from our machines: control, leverage, versatility, things like that.

Oersted showed that electricity could create magnetism. Could magnetism, conversely, create electricity? Now it was Faraday's turn. He expected to coax a current into flowing through a circuit if it were placed in the right way near a steady source of magnetism. He had a compass needle to indicate the presence of magnetic force, and he had a metering device, with a pointer, to indicate the presence of current in his coil. But the current didn't materialize. Then one day, as Faraday turned off his magnet, an electromagnet, he noticed the pointer of his electrical meter flicker into life briefly. Was that a current that had flowed in the circuit for an instant? Maybe this little transient occurrence observed out of the corner of his eye had been a mistake. Maybe it hadn't really happened. Many laboratory practitioners, impatient to get on with things and not sufficiently *alive* to new possibilities, would have ignored the flicker. The prepared mind, however, will be in a better position to seize on serendipity. Faraday was about to have a Eureka moment.

He didn't ignore the flicker. He repeated it, changed it, drew out its meaning. When he turned the magnet back on, the flicker at the meter came again but in the opposite direction. In the annals of science and technology, his was to be one of the greatest acts of deduction *and* induction. Faraday correctly deduced that it was a *change* in magnetism, not *steady* magnetism, that induced an electric current in a wire. This certainly wasn't what he'd been expecting, but what he had on his hands was a demonstrable form of magnetically inspired electricity. By inventing a way to move the magnet past the circuit or the circuit past the magnet in a consistent way—usually in some kind of cranked, spinning motion—a regular current could be produced.

Some discoveries are singular but not this one. On the other side of the Atlantic, in Albany, New York, Joseph Henry was making observations very similar to Faraday's. In fact, Henry seems to have glimpsed the flicker of induced electricity *before* Faraday but tardily published his results *after* Faraday did. And so, by the rules of scientific precedence, whenever the credit for magnetic induction is to be parceled out to a single name, the Englishman generally gets the nod. Henry was disappointed, but he didn't want to be seen making a fuss over the matter and apparently accepted the judgment gracefully.[3] Notwithstanding his failure to get a full share of the acclaim, Henry received more than a flicker of recognition later in life. He was a Princeton professor, a scientific adviser to President Abraham Lincoln, the first head of the Smithsonian Institution, and second head of the National Academy of Sciences. (Last but not least on his honor roll of attainments is the palpable fact that the book you hold in your hands at this moment bears on its spine, in the form of the publisher's imprint, the name of Joseph Henry.)

So, Faraday (and Henry) had brought something dramatically new into the electrical world. Electricity could be made not only through the chemical reactions in a battery but also with a new kind of machine, a dynamo, that converted a mechanical motion (a magnet mounted on a spindle revolving past a stationary coil) into electrical motion in a wire. There was a practical limit to the size of a battery you could make, but larger and larger dynamos producing larger and larger currents were a fathomable idea.

Early dynamos, alas, were inefficient, and Faraday's great insight did not revolutionize electrical affairs overnight. Batteries, however cumbersome, were deemed adequate for most applications, such as the telegraph. Indeed, electrical gadgets at first played only a minor role in the gathering industrial revolution, where steam engines and railroad technology predominated. Dynamos existed, wires carried current, and primitive electric motors were employed here and there, but still there was no grid.

Meanwhile, Thoreau's flicker of recognition, made at much the same time as Faraday's, was to perceive that machines, even as they help us be more efficient workers, can intrude on our lives in unexpected ways and can act as a barrier between us and other people or between us and nature. Thoreau's solution was to simplify his life, to clear away as many barriers as possible, and to create as much leisure time as possible for his preferred activities, namely reading books and writing his thoughts in a diary.

Faraday experimented with forces while Thoreau experimented with philosophy. Both could be poetic. Here they reveal their respective approaches to exploration, Faraday first:

> The philosopher . . . should not be biased by appearances; have no favorite hypothesis; be of no school; and in doctrine have no master. . . . Truth should be his primary object. If to these qualities be added industry, he may indeed hope to talk within the veil of the temple of nature.[4]

And Thoreau:

> A book should contain pure discoveries, glimpses of terra firma, though by shipwrecked mariners, and not the art of navigation of those who have never been out of sight of land.[5]

Neither Faraday nor Thoreau lived on into the age of the electrical grid. Big technology in their time meant railroads, or steel bridges, or the slow digital flow of information by telegraph. So how do Faraday's and Thoreau's very different recognitions play out nowadays? Where were Faraday and Thoreau on August 14, 2003?

First, Faraday. To start with, on that day millions of electric motors, all of them embracing some kind of Faraday induction, stopped performing. Without their electromagnetic sustenance, elevators, traffic, subways, and drinking water just stopped.

The flip side of this, the Thoreau side, was also in evidence on August 14. When asked to assess the effect of the power outage, many New Yorkers, after complaining about spoiled food or lost computer files, could express also a sort of joy. For instance, many spouses reported that they were glad for the chance to have an evening's worth of conversation rather than watching banal drama on television. Parents in many places appreciated the chance to do something different with the kids. One mom showed her daughters how to play jacks. "I would like this once a month," she said, giving in to the charm of the occasion.[6]

Pause for a moment. Would this very Thoreauvian idea work? Would the very goofy but delightful lark of turning off the grid really work? Every first Saturday of the month, say, there would be no electricity. Or if that scheme removes the thrill of unexpectedness, how about shutting down the grid randomly, one day a month chosen by lottery? They'd try it then people would start to complain; then like Prohibition they'd repeal it. No, this proposal wouldn't work.

Faraday, at least on this issue, is allowed to trump Thoreau. You can, nevertheless, appreciate the deliciously subversive side of the blackout. The August 14 event was a bit like the medieval Feast of Fools, the Yuletide holiday when in towns around Europe class distinctions were suspended, if only for a day, and masters and servants switched places, church observances were mocked, and revelry overruled solemnity. August 14 became an enforced holiday. Tolls on the New Jersey Turnpike were suspended, city buses in many places were suddenly free, workers were told not to bother coming to work the next day, civilians were in charge of traffic at many corners, and those filing for New York City tax extensions were given extra time. Late filers were asked to write the word "Blackout" front and center across their returns.

In Brooklyn a Dixieland-themed musical parade promenaded around the streets. The borough president, taking up a prominent post on the Brooklyn Bridge symbolically welcomed his tired constituents as they fled Manhattan on foot: "Hello, Brooklyn. You're home now," said his banner sign.[7] In what sense were Brooklynites finally "home" when they trudged across the bridge?

Forgetting the normal interborough rivalry and ordinary hometown pride, I suspect what the borough president meant when he welcomed residents back across to the Brooklyn side of the East River was that he was also inviting them back across a technological divide to a simpler, less complicated time. There was plenty of inconvenience mixed in with the nostalgia—you had to walk, not ride, to where you were going, and you would have to eat by candlelight—but in many cases this was equivalent to having a picnic. Providing it wasn't too inconvenient, the absence of electricity *was* welcome. At least for a night.

THE MEANING OF AIR CONDITIONING

Later we will have occasion to explore the engineering complexities that allow blackouts to happen, including the mishaps that set off the 2003 event. But as this gigantic de-electrification event was unfolding, what did people think was happening? With a guilty shrug, many suspected air conditioners were to blame. And rightly so, since air conditioners, requiring pitchforks of power, are among the hungriest of specimens in the zoo of socketed appliances kept in our homes. One of the most blessed inventions from the standpoint of ameliorating the summertime discomfort that lingers around concrete-encrusted cities or any hot place—humid or parched—an air conditioner is a member of the family of refrigeration devices whose job is to transport heat away from one place and dump it in another place.

Example. You remove heat from an interior room at 72 degrees Fahrenheit and push it outside, where it's already 96. Now, unwanted heat does not of itself flow uphill from cold to hot, but it will if you bring appropriate machinery to bear. Here's how it works. You give the unwanted interior heat to an intermediary substance, a pressurized refrigerant fluid, which carries the heat away to be off-loaded elsewhere, out in the backyard. It's like barging Manhattan's garbage out to Staten Island, which already has trash of its own.

The net effect of this process, of course, is to warm the outside air even more. But at least that warmth would be somewhere else. Sitting in your chilled living room watching television, you could care less about what it's like out back with the trash cans or in the stratosphere, where ozone is being dismembered by chlorofluorocarbons leaking from your machine (or did until the old-style refrigerant was banned). In this way electricity, in select areas of human existence, allows you to preempt nature.

Ah, but now we have a dilemma. Here is where the civics lesson begins. On the one hand, air conditioning must be good. By the flimsiest flick of your wrist, the turning of a dial makes a compressor speed up or slow down, obtaining for you any desired comfort level. But doesn't this separate you from nature? Isn't that bad? Aren't we supposed to stay close to nature? If the gods had meant the climate to be nice all the time, they would have made every place just like San Diego.

Wait, you say. Does that mean we have to suffer whatever weather comes along? Didn't the first caveman who sat by a fire, soaking up the warmth and roasting his meat, separate himself from cruel nature? Can you blame him?

The sermonette continues: Air conditioning is bad insofar as it makes a racket, scares away the birds, and cracks the moldings around window frames. Air conditioning is good because it makes a hot summer afternoon bearable. With artificial cooling, cities like Atlanta and Miami could aspire to being national cities, serious destinations for business and tourists alike, not just regional swamp encampments of the Deep South. Dallas in the summer might be the coldest major city owing to its corps of air conditioners, just as Moscow might be the warmest city in winter owing to its avid mobilization of furnace heat.

Air conditioning must be bad because it leaks bad chemicals into the sky, but who will tell the old lady, the one with the heart condition, or the mother of the infant whimpering with an ear infection that they should feel the summer full force? The sweat glands are the body's own private transmission grid for carrying surplus heat from the interior up to the surface, where it's released through an act of evaporation. This process can take its course, as nature intended, or it can be speeded up through the application of electric-powered air conditioning.

So the argument swings back and forth. No single air-conditioning unit can be blamed for the disaster of August 14 or traced to any particular case of skin cancer. And anyway the nasty ozone-busting chemicals aren't used anymore. No extra amount of heat wafting up from all the refrigerators in Boston can be blamed for the general retreat of the Greenland iceshelf, can it? How much electricity is too much? We don't yet know.

Was the electrical grid inevitable? No, not inevitable. The grid started small and grew and kept growing because it was better than other energy delivery systems. Stop to consider an ordinary activity, like reading a book after dark. For centuries, if you were fortunate enough to have a book in the first place, you would have lit an oil lamp or a candle. In the 19th century whale oil was popular, so popular that dedicated fleets of ships sailed the world over in search of the leviathan of the deep, a process immortalized by Herman Melville. Shortly thereafter whale fuel was overtaken by other fuels, by kerosene and gas, whose procurement, though by no means pleasant, entailed no ships splintered by whales, no voyages into unknown waters, and no Melville. Electric light came next. But when the electrical grid crashes, and we're welcomed back to Brooklyn, what do we do? We return to candles.

Did it have to be electricity? Does energy have to come in that form? What were the alternative late 19th-century energy sources? Steam, gas, water, and wind. What did the rival grids look like? If history had worked out differently and electricity had not been developed, could we still have modern appliances? Could we, for example, run an air-conditioner compressor on steam power? The short answer: We could. A steam engine can compete with the power of an electric motor, at least up to a point. Generating steam is just like generating electricity; indeed, electric power often starts out as steam. The steam pushes turbines, which convert the steam energy into electrical energy. For the moment, though, let's pretend electricity doesn't exist. We're talking about the steam grid. Each month you get a steam bill. You have steam outlets in your walls, the steam equivalent of circuit breakers, steam-driven telephones, steam lighting. Occasionally steam-outs occur, and then you get an apology from the president of the steam company.

How would a steamicity transmission and distribution work? Here you run into trouble. Steam can be sent through pipes, but it quickly cools, so it can't go far. Electricity can be sent hundreds of miles, steam only a few city blocks. Couldn't a small steam generator be stationed outside your window and all the other windows and the power transmitted by clattering belts and pulleys to the air conditioner? Not only would such a shrimpy steam plant be inefficient, but picture all the moving parts of such a Rube Goldberg contraption and the heat and the noise. Actually, that is exactly what many early mechanized factories *did* look like when motorized by steam or water power. Steam was tried. It had its chance. It worked until something better came along.

How about other 19th-century energy sources? Gas jets? Gas is good for cooking and heating and was at one time widely used as the elixir of illumination. It had—and still has—an efficient grid of pipes under city streets. But it gives off unpleasant fumes and flames and heat, and so it wouldn't make for an efficient or healthful mode of lighting or air conditioning. Wind power? Wind, when it's up, might produce some cooling. Unfortunately, wind blows hot and cold. On a hilltop it's stiff; in the valley it's limp. How about the power of running water? It might activate the compressor and a fan, but you'd have to live next to a river for this scheme to work.

No, for flexibility, density of power, ease in transmitting energy over long distances into tight corners, and with no leftover mess (at least not in your home), it's difficult to beat electricity as an all-purpose form of energy. Little wonder it's been so popular. All those fancy high-frequency electronic technologies—radios, computers, TVs—are impossible without the grid. These products now come in portable form, but at the end of the day, when you want to recharge the batteries, where do you go? You have to plug them into a wall socket somewhere.

THE ONCE AND FUTURE GRID

You might say we are caught in the grid as if it were a net, as indeed it is—hence our ambivalence. It liberates us and it ties us down. It enables and ensnares. It simplifies some things and makes other things more complicated.

Certainly, several other grids rival the electrical grid in pervasiveness. Drinking water spreads through an expensive underground network of pipes; urban gas for heating and cooking flows through understreet mains right next to electricity; sewerage lines, constituting the lowest of the grids, ferry unmentionable fluids and particulate matter away from homes and businesses; the telephone grid, draped overhead on poles, snakes up through conduits in the walls; and the railroad grid operates both intracity and intercity trains.

Even as the greatest engineering achievement of the 20th century, the electrical grid won't necessarily be around forever. Nothing is forever. Many formerly successful grids are now obsolete. For instance, the network of canals, once a major mover of freight through the countryside in the early 19th century, was largely overtaken by the steel logic of railroads. Roman aqueducts, once vital, are now archeological. Incan roads, the interstate highway of the mightiest north/south-oriented empire in history were put out of business by a few dozen marauding conquistadors. A network of feeding stations for post horses kept intercity travel moving. But they're gone now since horse power isn't needed in an age of engines. In a few places, hydraulic utilities sent pressurized water or oil to selected locations to turn motors or lift elevators. They're gone now. Ice: Frozen water doesn't sound like a grid, and yet there was a time when ice carts moved regularly up and down crowded city streets. Not to chill drinks. Ice preserved foods for another day. But since the arrival of refrigerator freezers the iceman cometh no more. The list of vanished grids is long.

The electrical grid, made of metal, is not a living thing, but an impartial observer from Mars might suppose that it comes close to being alive. Look at the manifest gridness of the grid. Like the human body, the electrical grid possesses a sort of nervous system that both senses and actuates. The grid constantly samples its local environment (for example, it affirms that current is flowing through appropriate wires) and sends appropriate commands to outlying sectors. The grid has an equivalent to the endocrine system—electrical instead of chemical—in that it constantly executes fine adjustments of vital parameters (voltage or frequency levels) as they are needed to maintain proper hormonal balance. The grid has a counterpart immune system (consisting of, for example, protective circuit breakers and relief valves) for the self-healing of various disorders. On August 14 dozens of electricity generators, sensing trouble, turned themselves off in a sort of collective allergic reaction to the threat of overload.

The grid possesses a digestive mechanism for the consumption of primary energy (coal, say) and its transformation into useful work. It has an excretion process for off-loading waste products such as smoke and spent cooling water. It has a skeletal network of supportive scaffolding (such as high-voltage towers). Perhaps the most obvious anatomy analogy between the living animal and the electrical grid is the crucial and central position of a heartlike dynamo (or set of dynamos working together) propelling a circulating, energizing fluidity out and through and around an extensive vascular web of arteries and capillaries.

The numerous organs and habits of mammals, tuned by the relentless forces of evolutionary biology, have been optimizing for millions of years. By comparison, the subcomponents of the power grid, engineered by human ingenuity, have been around for a scant dozen decades. This is far less than the lifetime of a single Sequoia, so it's too early to make predictions about the longevity of the grid in planetary or cosmic terms.

Most students of technology would bet the electrical grid will be around for a long time. Yet on August 14, 2003, a considerable chunk of the North American grid was out sick. The roster of places that lost power included eight U.S. states and two Canadian provinces. The number of people who were deprived of power numbered about 50 million. Business losses—spoiled meat, stock transactions suspended, manufacturing wasted or idled—added up to billions of dollars. The event was the largest blackout in history as measured in terms of power turned off—as much as 70,000 megawatts.[8]

How big is a megawatt? How much energy is that? How impressed should we be? First of all, a watt is not a unit of energy but a rate of energy use: energy *per second.* A million watts, or one megawatt, is the power consumed by roughly a thousand households. It's the equivalent of 10,000 100-watt lightbulbs. It so happens that an average human body exudes about the same heat as that bulb, so a megawatt of electricity can be compared, in terms of power, to the rate of body heat emitted by 10,000 people, enough to crowd a place like Madison Square Garden for a basketball game.

On August 14 then, the amount of power that vanished across the blackout region, in terms of perspiring human bodies, would have been 70,000 megawatts times the 10,000 perspiring people per megawatt. That comes to 700 *million* people. In other words, the power surrendered on August 14, in the space of 12 seconds, was equivalent to the body heat given out by more than half a billion people. Picture all those fans packed elbow to elbow, leaning forward in their seats, the championship game decided as the ball goes through the hoop at the final buzzer, all that moist breath exhaled, that magnificent aura of bodily warmth rising out of the stands at the climactic moment. And then it all went away as if with the flip of a single switch.

Thoreau might have been gratified. For sure, the Dark Sky Association, dedicated to clear skies and improving astronomical viewing, was gratified.[9] The great electrical mishap and the withdrawal of energy from thousands of fluorescent bulbs gave the association a chance to make its point. There is a universe out there, separate from and indifferent to human concerns. The dark sky was a bonanza for those who wanted to see this indifferent heavenly realm. Stars that hadn't been seen in years were now visible. Mars was particularly fetching. Shooting stars were seen. Gazing on the Big Dipper made some grown men and women as giddy as children. There were even unconfirmed reports that some individuals, within the borders of New York City, had actually been able to perceive that grand armada of stars that make up our home galaxy—the Milky Way. You could imagine some celestial borough president welcoming you, the night-sky viewer, home again, to a sight you hadn't seen for years. Maybe not ever.

The 2003 blackout was inadvertently an immense chemical experiment. When the electricity stopped, so did the pollution coming from fossil-fired turbogenerators in the Ohio Valley. The before and after air quality levels were dramatically different. On August 15, only 24 hours after the blackout, sulfur dioxide was down 90 percent, ozone was down 50 percent, and soot particles were down 70 percent from normal conditions in the same area—"normal" meaning polluted.[10] Less electricity, in this case, meant cleaner air. The atmosphere could in effect say to lungs everywhere, "Welcome Home."

Søren Kierkegaard said that life is understood backward but lived forward. The same is true of technology. We often build machines and only later learn what they're for. It's time now to encounter the grid in the forward direction, in the manner in which it was constructed, from its infancy to the present day, when it is still undergoing turbulent growth.

The proper story of the grid must begin with Thomas Edison. Please imagine that the year is 1880. He has already invented and patented marvelous electrical devices such as the phonograph. Next, he will do something even more marvelous: invent the *system* into which all those other devices would fit. For us the grid has been around for more than a century, but for Edison it was new. He had to bring it into existence. Indeed, at first it existed only in his head, and on greasy blueprints, and in the form of some wires slung behind his laboratory. Edison was about to do what Ben Franklin could not do. He was about to put lightning into a bottle and sell it. Electricity would be available as if it were water pouring from a tap. Because of him electricity was about to burst forth in the city. It was the morning of the grid.

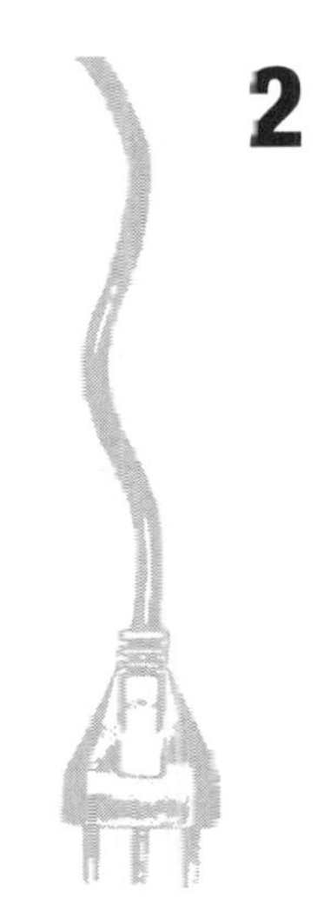

2

GRID GENESIS

It was the morning of the grid, one o'clock in the morning, in fact. That hour was all that her busy schedule would allow. He was said to be the most famous man in America and she the most famous woman in France. He was dedicated, efficient, indefatigable, ingenious, while she was, to distill the litany of superlatives down to a single word, simply "divine." On a chill night, after her final Broadway performance, Sarah Bernhardt crossed the Hudson River for a midnight rendezvous with Thomas Edison. If she was an enchantress, he was a wizard, and for her arrival at the lab he set out to dazzle her with the best electricity at his disposal. The first thing she would see would be his forest of lamps against the darkness.

He spoke no French, she no English, so all that was said between them necessarily passed through an assistant present to perform instantaneous translation. Another assistant was assigned to prevent her layered skirts from becoming entangled in the apparatus as the couple swept from room to room. At certain moments they even held hands. She was the most extraordinary woman he had ever met. To her, his sagacity made him look irresistibly like Napoleon. His expression had been imperial, confident, and skeptical. She had come determined to conquer him and had succeeded.

Before leaving, the Divine Sarah recorded some lines from *Phedre* into Edison's renowned phonograph, all of three years old, achieving what the 17th-century French playwright Jean Racine would never have expected—his immortal verse turned into immortal sounds by the inscribing of shallow grooves on a foil cylinder. Joyfully reciprocating, Edison sang a few stanzas of "Yankee Doodle Dandy," making Mademoiselle laugh. Everything about the evening was incandescent.[1]

A few weeks later, in these same rooms, Edison hosted a much more consequential visitation, one that would properly launch the electrical power business in America.

SEEING THE LIGHT

Thomas Edison's story is of course indispensable to any account of electricity. He is Moses, the lawgiver of grids. He not only brought the first big-city electrical grid into existence but practically invented the concept of gridness, and he did it through (to use his formula) 10 percent inspiration and 90 percent perspiration. Behind this simple equation lies decades of 18-hour workdays, long nights in the laboratory searching for technical solutions to resistant technical problems, and tedious expeditions among skeptical bankers for seed money.

Edison's inventing life had been bound up with electrical products: an improved form of the telegraph, allowing four separate signals to pass on the same line; a rudimentary "speaking telegraph," or telephone; and the phonograph. But until 1878 he hadn't been especially interested in electricity for lighting or the generation of electricity itself. On September 8 his attitude changed. Many were the visits that others made to Edison's lab, but on this day Edison visited the factory of William Wallace, a Connecticut manufacturer. What Edison saw—brilliant arc lights powered by a noisy dynamo—set his mind racing. In the arc process a metamorphosis takes place: a supply of energy in electric form flows from the dynamo to an electrode. Then the electricity, released from the chrysalis of its wire, leaps the gap from one electrode to another, tracing out a miniature lightning bolt. Some of the energy becomes a fierce whiteness that hurries away at the speed of light. It *is* light, and it can do nothing but escape in all directions.

On September 8 Edison saw the light. This was only the first in a chain of stimulations, however. Inside Edison's head the pulse of light was converted into a neurological flash, and this internal illumination, in the marvelous mental reconfiguring we call inspiration, became a vision of what would later become the electrical grid.

On that September day it came to Edison that he could do it better, cheaper, bigger.[2] He would invent not merely a better lamp, he would invent a system. The lamps would be mass produced and would come in many forms. The system of energy and lamp and fixture would be served by a centralized power source and a branching distribution grid. That would be just the start. The same wires would bring energy to other machines. This grandiose scheme was conceived by, and could only be implemented by, one man, himself—Thomas Alva Edison. And this was not the silver-haired septuagenarian of later years. At this point Edison was 31, a man in the pink of health.

The work at Menlo Park, Edison's invention factory hideaway in the Arcadian New Jersey countryside, would now be devoted to embodying the vision into an actual grid made of copper wire. That grid, which in its entirety today stretches around the world, started small as the experimental tinkering of master mechanics struggling at a workbench. At Edison's lab and at those of his competitors elsewhere in the United States and in Europe (Edison did not have a monopoly on brilliant electrical inspiration) implementation was slow. The time for 1,000-mile transmission lines and multistate blackouts would come decades later. Presently the key was finding the right recipe for a half-inch-long bulb filament that would take the heat and render an effusion of light for months.[3] Edison also had to ponder the effective formula for an insulator wrap that would protectively sheathe cables destined to lie in the damp earth for decades. He was doing the gruntwork needed for later greatness.

THE VOLTS OF NEW YORK

Edison's research blossomed in New Jersey, but the Edison grid was to be deployed in New York, for that was where the potential customers lived and where the financial investors had their affairs. And just as athletes of today often prefer to play ball amid the publicity glitz associated with a big-city media market, so Edison sought the notoriety that would flow from an engineering success carried out audaciously in the heart of the business capital of America.

What happened on the evening of December 20, 1880, was a prelude to the grid. After dark on that night, up and down a considerable stretch of Broadway, gentlemen were able to read newspapers on the sidewalks. It was that bright. What had happened was the startling debut of the first official large-scale outdoor electrical lighting in New York City. In arc lamps far overhead, impressive currents made their reckless jump from one electrode to another, powered by a dedicated wire. It was a wonder to behold, artificial moons safely out of the way on top of their poles, casting brilliant pools of light on the ground below.[4]

This wonder represented the high technology of the day, but it was not Edison's handiwork. Charles Brush, inventor extraordinaire and astute businessman, had carried the arc-light enterprise to a state of readiness and efficiency over years of trial and error. He was starting to move ahead of his rivals and had orders from numerous cities for his system of arcs, which would be energized by powerful central generators. Seemingly, urban activity in many places would be extended and improved by the well-placed presence of these artificial, vested lightning bolts.

And yet this day did not belong to Brush but to Edison. At the very moment of its inauguration, the arc-light grid was being challenged by activity elsewhere. Across the river on the same evening as Charles Brush's triumph, something more important was happening.

At the last minute the mayor of New York, Edward Cooper, regretfully declined Edison's invitation, but on that night a sizable contingent of city aldermen, accompanied by numerous gentlemen of the press, crossed over to New Jersey to see for themselves the marvelous gadgets and learn how the wizard of Menlo Park planned to electrify their metropolis. The city fathers probably didn't need much convincing, but they did want to extract from Edison a fee for gaining permission to tear up the streets of lower Manhattan, and it was Edison's job to avoid paying that fee. He did this by impressing the pants off them. After leading the men around the labs (no skirts, this time, to get caught in the apparatus). Edison guided them to a window where, with the lifting of a lever, he lit his arbor of outdoor lights, which had produced, as it had with Madame Bernhardt, a ripple of delight and amazement in the onlookers. This display was but a junior version of what he had in mind for their city. The Menlo Park network was only a toy grid.

Thomas Edison was already world famous because of his phonograph and efficient mass-producible lightbulb, but an even greater accomplishment would be the installation, maintenance, and growth of a system, a centralized system, of domestic electricity in the city. First hundreds, then thousands, then millions would benefit. Those lucky enough to be alive at this time and those in states unborn and speaking with accents yet unknown were destined to live in an electrified time because of this extraordinary scheme. Surely the gentlemen of the city would want to share in this historic undertaking.

The aldermen, interested and impressed but perhaps fatigued by Edison's exhaustive reports, were next led to a dimly lit hall. Then came the theatrical coup. With the throwing of yet another lever, the room was flooded with lights from every side, revealing a sumptuous feast, catered by the popular New York restaurant Delmonico's. The evening finished up with champagne, Cuban cigars, and effusive toasts to Edison's genius. (Picture something like the banquet scene from *Citizen Kane* but without the dancing girls.) The men who ran New York City had come to the lair of the wizard expecting to be amazed and were not disappointed. As one of the newspapers summed up the aldermanic evening, "They came, they saw, they marveled." Edison got his permit to dig.

PEARL STREET

The mountain had come to Mohammed. The powerful New Yorkers had crossed the waters and come to New Jersey, and now the wizard would return the favor. He moved his troops and all their equipment to the heart of Manhattan to be near the action. Supreme allied headquarters for the Edison Electric Illuminating Company of New York was a fancy building on Fifth Avenue just south of 14th Street. Other shops and factories for support operations opened up all over town. Here is where the matériel of the grid—circuit breakers, bulbs, fixtures, conduits, and above all wiring—would come into being. But the venue of interest to us now, and arguably the most famous address in the annals of the electrical grid, would be 257 Pearl Street, in the Wall Street area just below the foot of the almost-finished Brooklyn Bridge. It was from here that electricity would soon be bestowed on customers over a square-mile zone and eventually, in a larger sense, on people all over the world.

The great bridge linking the cities of New York and Brooklyn (the boroughs hadn't yet been amalgamated into the larger urban enclave we know today) is justly exalted as a historic engineering feat, a one-of-a-kind masterpiece. By comparison, the grid going up (or rather down since it was mostly underground) in the bridge's shadow was to be repeated in every big city in the world in pretty much the same form. The Brooklyn Bridge would change the flow of people and goods in New York City and its environs in a big way, but the warren of wires being sunk in the dirt would transform the city even more profoundly.

It would be good to say what the Pearl Street operation was *not.* It was not the first time a building had been lit inside entirely by electricity; Edison's own headquarters at 65 Fifth Avenue held that claim. Nor was this the first time Edison had served up electricity to customers; several isolated generators of his were functioning in the homes of New York's wealthiest, including Edison's chief financial backer, J. Pierpont Morgan.

Being self-sufficient when it comes to electricity—having a power plant in your backyard, that is—sounds desirable, but in practice was a headache. The generator often failed to perform. When it did work, it emitted fumes and noise, causing those living nearby to complain. Also the wiring and insulation in those early days left much to be desired. The result was often singed carpeting and browned furniture. Although Mr. Morgan expected the electrical problems in his home to be put right, he was apparently patient with the repairs and was reportedly fascinated to be part of a historic experiment in the domestic use of electricity.

The Pearl Street operation would not be the first time the streets of New York City were illuminated by electricity; those arc lamps, ushered in on the night of the Edison banquet, had served admirably in splashing buckets of light up and down Broadway. It was not even the first time power had been supplied to multiple customers from a centralized station; after all, Edison had lit up part of Menlo Park as a sort of welcome mat for those like Madame Bernhardt or the New York aldermen or the many who had bought train tickets just to gawk at the Edison compound by night lit up like a Christmas tree as they rode past.[5] Then, too, in January 1882 Edison had set up a station in London and sent electricity to establishments along the Holborn Viaduct with wires strung overhead.

What the Pearl Street system *did* represent was the beginning of large-scale, centralized electrical delivery to a diverse clientele for what would soon be a variety of uses. This was no longer an experiment conducted in dark laboratories but the beginning of regular service. Your electricity would be there in the morning and the evening. It would not come from a battery or a laboring generator unit in your basement. Instead current came unseen from some other place, a dynamo first blocks, then miles, away. The Edison system would make electricity more intimate by separating the user from the source. Just as water from a distant lake and controlled by a handy tap had become routine, so too electricity would seem to become just another everyday flowing fluid.

True, this nonchalant familiarity bred of long usage would, at times, complicate the relationship between humans and high voltage. Electricity was not just a product off the shelf but a dynamic process, a potentially dangerous substance if touched, bearing a normally explosive form of lightning-like energy tamed and constrained to a wire. It was a continuous enabler of activity, a *presence* in everyone's life. Unless you are on a mountaintop or a forested footpath you're probably, at this moment, within a few meters of something electrical.

The Edison enterprise in New Jersey had started as an invention sanctuary, a retreat for a monastic order of creative tinkerers. Now the plans called for a more vigorous proselytizing mode of action. Like the Jesuits, the Edison men, trained in secret and inculcated in a fervent catechism, were called on to journey out into the world, first to New York and then farther afield, to earnestly spread their gospel of electricity. Their first parish would be District 1 in New York City. Others would follow. Some of Edison's most trusted lieutenants were dispatched abroad as missionaries to London, Paris, and Milan. Berlin wanted to build its own Pearl Street and asked for Edison's very blueprints.[6]

Edison had always been a busy man, but now his schedule became frantic. For the greatest inventor in U.S. history, his output of patents would peak in these years of grid construction.[7] He was an inventor, but now perforce he was also a manufacturer, a mass producer who extruded wire by the mile and assembled massive machines that combined and controlled all the fiercest man-made energetics then known: fire (in coal-fed boilers), steam (for driving pistons), and electricity (cranked out by the best dynamos in the world).

For this commercial reformation to take place it had to work in New York first—and so far he was behind schedule and over budget. The labor of building the actual grid had become a grind. The financial backers had long been patient, but now they were eager to get things going. Where was the revolution Edison had promised? When would the currents flow forth from 257 Pearl Street?

DIVIDING THE LIGHT

At last it was done. On September 4, 1882, Thomas Edison had a full day ahead of him. He took off his coat and collar but kept on his white derby hat. The testing was over, the machines were installed, the wires had been laid, and now it was time to energize the system with the throw of a switch. Edison, beginning the day with his men in the field, was making a final check of the cables. He relished the attention he got from journalists and appearing at street level helped burnish his folksy image. It was not all sham, though, since Edison did seem genuinely to like being where the hard work was. That's what had brought him into the ditches by the side of the road with the Irish laborers on the morning of the big day.

The grid had been tried out section by section and not without problems. Approval from the fire insurance underwriters had been slow in coming—and for good reason. At one street corner, for instance, faulty insulation had allowed some of the current to come to the surface, giving passing horses a nasty jolt whenever they trod in a particular puddle of water. Most of the grid was underground. Copper had been mined from the earth, drawn into wires, and now Edison was putting it back into the ground. It was expensive to do this, but Edison's grid—unlike the competitors' overhead welter of wires perpetrated by numerous small-scale voltage vendors for delivering telegraph messages, burglar alarms, fire calls, and telegraph signals—would survive well into the next century. (On a Wall Street stroll I once happened to see some of Edison's underground brick vaults, exposed to the air for the first time in 125 years because of road repairs.)

The business end of the enterprise consisted of a mechanized work gang of six huge steam-generator sets, each capable of furnishing juice for 1,200 lamps, making them the new world heavyweight dynamo champs. These behemoths, each about the size of a reclining elephant, and consequently known as "Jumbos" after the famous performing elephant in P. T. Barnum's circus. Very soon, in fact, Edison's Jumbos would be performing in the Paris Opera House.

Edison, normally quite free with the press, had suddenly gone shy. On this day he declined to speak with reporters until the deed had been done. Only then, he said, when scoffers had been silenced, would he give his opinion of the significant events about to take place. With everything in place, Edison donned his fancy duds and went over to the offices of J. P. Morgan.

The time for gritted teeth was fast approaching. So far, and for too long, the investment had been all outflow, with no income. Therefore, although there was much mirth in the room, there was also solemn concern. A half million dollars had been spent up to this moment and not a morsel of dividend had come back to the investors, some of whom were at Edison's elbow now waiting to see the advertised wizardry.

Edison was the Odysseus of volts—clever, resourceful, brave, and persistent. Edison had done all he could—invented the devices, tested the ideas, hired the engineers, worked prodigious hours, organized efficient means of manufacture, wrested municipal approval from skeptical politicians, and accomplished the installation. He had built a city within the city, a web of hidden wires. The time had come to throw the switch in Morgan's office. Everyone in the room, especially those who had backed the scheme, needed the electricity to arrive at the designated lamp sockets. Would the electricity, and shortly thereafter the profits, flow as promised?

The answer to the first was a resounding yes, while the answer to the second would be . . . not for several years. At 3 p.m., Edison melodramatically threw the lever, and the bulbs in Morgan's suite successfully energized.[8] The gods had smiled. Over on Pearl Street, an instant before the climactic moment, the sturdy copper circuits had faithfully borne power to 400 lamps scattered around District 1, a region bounded by Pearl, Nassau, Spruce, and Wall streets. Jumbo Number 9, performing solo that day, rendered blazing light by converting the rotary motion of a shaft set spinning by a steam engine, which in turn used the chemical energy stored in a few tons of coal compressed deep in the earth hundreds of millions of years before in the Carboniferous Period from beds of decaying green plants, which in life had been sustained by the photosynthetic process powered by ambient daylight sent hither from the photosphere of the sun.

Now that the grid revolution was launched, what did it look like? Probably not like a revolution. Gas lighting had been around for decades and had no intention of surrendering. As for electric illumination, arc lights had been up for almost two years, allowing New Yorkers to walk at night easily up and down the larger avenues. They wouldn't be amazed by the sheer presence of massed brightness. It turned out, however, that even those jaded by light were impressed by Edison's lamps, by the mellow glow that bathed rooms as darkness started to close in. You could argue that Edison's filaments weren't as brilliant as the arcs, but *that* was the point. Edison had brought electric light indoors, into shops and parlors. When you sat down to read, you didn't want a miniature sun over your shoulder.

In one practical sense, it had been harder to supply low and steady light in a small glass enclosure than it had been to produce an arc, which was, after all, a stunted portion of bottled lightning allowed to come out of its bottle. Edison changed this. His name comes first on the list of illumination all-stars because he had usefully *divided* the light, had brought the fireball down from its tower, tamed it, packaged it, domesticated it. When you went to bed, perchance the last thing you did before going to dreamland was to switch off an incandescent bulb. In the morning, you could turn it back on. Edison and his grid were there at the ready.

The electricity revolution was still in its early stage, the Battle of Concord phase. The shot had been heard around the world. New Yorkers, if they had enough money, could have an electric current from a central supply source diverted into their homes or offices. Edison had brought electricity indoors, where its chief competition was not going to be arc light but gaslight. This battle of combustion versus incandescence would be fought for years, even decades. It took this long because electricity was then and for years to come too expensive for most households. The grid was definitely for upper-class customers only.

On the first day of electrical service, the wiring had threaded its way through 400 lamps. A month later this number had doubled, and two months after that it had tripled. That's where the electricity went: lamps. There weren't yet many other uses. You contracted for lamps. When a bulb blew out, a man with his tools came to your house and replaced the bad bulb with a good one. The electricity itself was free, at least at first. This introductory honeymoon period didn't last for long. On January 18, 1883, the Ansonia Brass and Copper Company got the first bill, for $50.40.[9]

The lamps were made to be uniform, but some, such as those in Mr. Morgan's office, were more equal than others. More significant still, for spreading the word about bulbs and augmenting the Edison legend, were the lamps plugged in at the *New York Times* building, which was then not in Times Square but downtown in District 1. In fact, among first-year customers, the *Times* had the biggest share of lamps—300. The smallest customer, the National Fire Insurance Company, subscribed for only three bulbs.[10]

The *Times* reporter, like many others, preferred the warm, even glow of the filament bulbs to the glare of the high-power arcs. In his report in the paper the day after the big switch-on, he describes both the ethos of the journalist's craft and the sheer curiosity felt by anyone who saw the light of day extended into the hours normally reserved by nature for darkness:

> The electrical lamps in the *Times* Building were as thoroughly tested last evening as any light could be tested in a single evening, and tested by men who have battered their eyes sufficiently by years of night work to know the good and bad points of a lamp, and the decision was definitely in favor of the Edison electric lamp, as against gas. One night is a brief period in which to judge the merits or demerits of a new system of lighting, but as far as it had been tested in the *Times* office the Edison electric light has proved in every way satisfactory. When the composing rooms, the press rooms, and the other parts of the *Times* Building are provided with these lamps there will be from 300 to 400 of them in operation in the building, enough to make every corner of it as bright as day.[11]

New York was Edison's headquarters, but his business ventures were global. The same year as the Pearl Street debut, Edison light systems illuminated a theater in Santiago, a railway station in Strasbourg, a post office in Budapest, the Grand Foyer of the La Scala Opera House in Milan (where the Edison power plant was the largest in Europe), the Wilhelmstrasse in Berlin, a café in Havana, and the czar's coronation in Moscow.[12] Everywhere Edison went there would be competition, but for a magic moment he had a jump on his rivals. He was on top of the world.

Back in New York, the grid was already falling behind. The first order of business for any grid is supplying power as it is needed. At night, people naturally use more light and the system operator had learned how to keep up with the increased need for electricity by having the stokers shovel coal faster into the big boilers in order to make more steam to handle the greater burden on the generator. At first this job of instantaneously anticipating customers' needs was easy. Increased darkness called for more light, which called for more electricity, which called for more coal. The art of meeting the electrical load became so well tuned that even the falling darkness from a midafternoon thunderstorm, as relayed down to the boiler room by a boy stationed on the roof to keep watch of western skies, could be countered neatly by a calibrated extra shovelful of coal flung into the voracious furnace.

How do you fold new customers into the scheme? As demand grows, you could add an extra Jumbo to the generator lineup. Beyond a certain point, though, you would need a whole new power station, and that takes time to build. Very quickly the Pearl Street system became oversubscribed. Additional "applications" for lamps were being turned down until such time as spare current became available. The company would, if you liked, put your name on a waiting list.[13]

Considering it was the first grid, the Pearl Street network worked surprisingly well without interruptions for seven years. New York City's first blackout occurred on January 2, 1890, when a fire broke out at the power house, possibly because of overheating in some upstairs insulation.[14] The Pearl Street building was gutted, and the only salvageable piece of equipment was the stalwart Jumbo 9, the star performer back on opening day. (Actually, this beast of burden was destined to participate in several more notable grid events, as you will see.)

The company was chagrined over the fire, and messengers were sent to users informing them of the problem, asking for their patience, and assuring them that service would resume quickly.[15] Power from an alternative plant (already Edison was colonizing other parts of the city) soon arrived, and Pearl Street itself was quickly back in operation. In the meantime, could the customers please refrain from using unnecessary lamps? So, besides being New York City's first blackout, this occasion had inspired the first electrical energy conservation campaign.

THE HUNGARIAN REVOLUTION

Nikola Tesla would eventually be precisely in the right place at the right time with the right idea. Moving about within the Austro-Hungarian empire in search of opportunity, Tesla was far from the action in New York, Berlin, and London. Born in what is now Croatia, he went for technical training to the Austrian city of Graz and then the Bohemian city of Prague. Tesla was an intense fellow, severely limiting his leisure activities and even his sleep in order to devote himself to the study of scientific, mathematical, and engineering subjects—not that he was a recluse. He was always a snazzy dresser, liked to recite poetry, and was one of those people who are almost frighteningly brilliant and focused. Under different circumstances he might have become a pianist or a professor of mathematics. Actually, what he loved more than anything else was electrical devices. He wanted to fix them and alter them to perform new tasks. Like Edison, he wanted to help mankind by extending the reach of electricity.

One can't write about Tesla without mentioning visionary thinking. Mostly daydreams are of short duration and small use, even to those who dream them. But to someone of sharp intelligence and perseverance, such as Tesla, a high-precision daydream could blossom into something substantial, even historical. We've seen how Edison's visionary moment (perhaps one of many such moments)—the one in which he grasped the notion of a complete electrical *system*—led to a coherent grid of machines and lamps nourished by a rain forest of wire.

Tesla's mesmerizing moment happened while walking through a park in Budapest. Ever prone to musing about machines, Tesla on this day pleased himself by solving a problem he had wrestled with for some time. In this daydream the design for a motor began to materialize. As an imaginative exercise he could start the machines, stop them, take note of minor problems here and there, noticing, say, some fatigue in the metal or a need for some rewiring. What emerged was a grand, harmonious, efficient, operating grid system, a better system than Edison's.[16] Who needed a lab when, with creative thinking, all the required tests could be carried out as an act of imagination? But even Tesla the visionary recognized that for his Budapest grid to be transformed from mental wiring into copper wiring, his material prospects would have to change. Actually, Tesla *had* built a real prototype motor, and it worked exactly like the idealized Platonic motor in his head.[17] Now it was time to make the machine available to the rest of the world.

Edison and Tesla: what a pair. Edison lived in New York. He had a great reputation and had realized his electrical vision. Tesla lived in the far corner of a declining European empire. Edison, in order to create his grid, had merely crossed the Hudson from New Jersey to Manhattan. For Tesla to realize his grid, it would be necessary to cross a more substantial body of water.

From Budapest, Tesla made his way to Paris where he got a job with the Compagnie Continentale Edison, which dispatched him on troubleshooting missions to the infant grids in several cities, where Tesla made himself indispensable as a fixer and redesigner of machines. The particulars of his Parisian employment, in the end, did not live up to his expectations, and so the impatient and resourceful Tesla did what millions of other young Europeans were doing in the 1880s. He set off for the promised land of America.

It is not surprising to learn that a man whose many waking thoughts were spent in a universe of pretend machines powered by pretend electricity would have his pocket picked and his luggage snatched soon after arrival. He arrived at the southern end of Manhattan destitute but with an appointment to work for Edison. This arrangement was to be invigorating, testy, and short. Both men were industrious inventors. Both regularly worked 15- or 18-hour days. Where they came into collision and why they parted after less than a year was over a thing that might seem small but would loom very large in the history of technology. Their great dispute was over the way electricity should move through wires. Edison promoted direct current (DC), current flowing continuously in one direction, like blood being pumped away from the heart, while Tesla advocated the use of alternating current (AC), current that flows first in one direction and then back in the other. AC electricity comes and goes like a tide, not twice a day like ocean tides, but rather many times every second.

Edison felt that anyone who preferred AC, a current that couldn't make up its mind, was a fool. When Edison and Tesla separated in 1886, AC was still a novelty, hardly more than an idea, seemingly headed for niche applications only. By contrast, DC had proved itself. Already the Pearl Street electric district had as many customers as it could handle. Why would anyone, customer or utility vice president, want to meddle with a successful formula? Why build or rebuild power stations or wire machines all over again? Did the wheel need reinventing? The historical answer to this last question was yes.

After leaving Edison's employ, Tesla fell much lower in the world—he went bankrupt in a business venture and was obliged, for a time, to work as a manual laborer, even digging ditches—but he was never far from a laboratory. He never gave up on that Budapest vision.

To appreciate what Tesla did next, how his fortunes turned, and how electricity became strategically useful in the workings of everyday life, let's look at the phenomenon at a more intimate level. A close-up discussion of electricity could have intruded at any of several points in the story so far: when Ben Franklin brought lightning down from the sky on a kite string, when Volta pushed electricity through a circuit with a battery, when Faraday (and Henry) saw that faint flicker of induction from across the room, or when Edison dispatched energy out beneath the streets of Manhattan. Nevertheless, *now* is the proper time to be initiated into the lore of electricity because of what Nikola Tesla had in mind.

THE MYSTERY OF ELECTRICITY REVEALED

Let's start with the homely but practical comparison between electricity flowing down a copper wire and water flowing down a brass pipe. What flows in the electrical grid are tiny charged particles—electrons. What flows through the plumbing grid are tiny water molecules. Water flow can increase if you widen the pipe or increase the pressure behind the water. Correspondingly, electric flow can increase if you use a thicker wire or increase the voltage, the force that impels the electrons through the wire; indeed, in the early days voltage was often referred to loosely as electrical pressure. So far, so good. The visual comparison is pretty direct: moving water molecules and moving charges. The popularizer of electricity at this point might rest content. But to get nearer to the reality of electricity, a more subtle comparison will be needed.

Picture a freight train preparing to start up from a dead halt. At the front end of the train the engine slowly starts up and, engaging the first car, gives it an abrupt yank. This car, in turn, engages the second car; there is an audible clank as the slack between the two cars is taken up. Then successively the slack is taken up with car after car after car, each time with a resounding sharp jerk. This jerk moves quite fast, propagating down the length of the train like a rifle shot echoing in a canyon as the cars, one after the other, get entrained. This is what electricity is like.

Look again at the train. There are two movements at work. The engine and all the cars are now moving slowly, at first at no more than a walking speed, but the moving linkage, the rifle shot connection between the cars, has moved much faster. This is what electricity is like. Electricity is really *two* things: the moving of electric charges and the moving of electrical *energy* in the form of the linkage between the charges. These two things, the charges and the energy, move at different speeds. The rolling linkage zipping down a wire travels at nearly the speed of light, whereas individual electrons are moving at only a very small fraction of that speed.[18]

Electricity is not precisely the charges but rather the thing that gets passed on by the charges. This is true whether we're talking about a circuit that loops around the inside of a flashlight in your hand or the circuit that goes all the way from the hydroelectric plant at Niagara down to a customer in Brooklyn and back again, a round-trip of 800 miles. Electricity is not merely a river of charge but the rifle shot of linkages sent along by the charges.

In this book the word *electricity* might sometimes refer to the flow of electric charges, but more often it will refer to the transfer of electric energy since the energy is what actually makes machines run. So, in summary, *electric charges* are the tiny particles, electrons, that move through a wire; *electric current* is the flow of electric charges (so many charges coming past per second); *electric energy* is the energy (the potentiality for actuating a machine or lightbulb) transmitted by the connections among the charges; and *electric power* is the amount of energy sent or used per second.

We're nearly done. One more strategic concept was needed before the discoveries of Franklin and Volta could be used in a practical way, and this was the fruitful idea of *electric force field*. Faraday provided the early insights, but more thinking was needed. Normally we describe "force" as being a push-pull kind of influence. You push a wheelbarrow. A strong wind pushes you. An elevator takes you upward. In these examples the forces are exerted between objects in contact with each other. But what about that linkage between the charges? It's called electric force. Where is it? What is it?

Isaac Newton provided the first major scientific explanation of an apparently noncontact force. When asked what gravity was exactly, he said that he didn't know. However, he was able to offer something just as good and maybe better. He explained what gravity *did*. He described gravitational force with equations so valuable that one could use them to compute in detail the behavior of cannonballs, comets, and (centuries later) space capsules.

Finally, *fields.* Abstract in description but very real in their effect, fields are the key to vivifying the matter of force in our minds, the crucial advance needed to push from Faraday to Tesla and arrive at the electric and radio world we now inhabit. What goes between the earth and moon, to go back to the Newtonian example, is a gravity field, a sort of connection radiated or flung across empty space. Actually, space wouldn't be exactly empty since it would be filled with the gravitational *fields.*[19]

And now the big payoff. The concept of fields applies as well to depicting how electric and magnetic forces operate between or among charged objects or currents. You may have seen iron filings sprinkled on a sheet of paper held above a bar magnet organize themselves into a characteristic oblong pattern. What is happening is that the filings, each acting like a tiny compass needle, are aligning themselves along the lines of force spewed forth by the magnet. The filings help visualize where the magnetic *field*—the streamers of magnetic influence—is deployed in the vicinity of the magnet. The field is a shorthand for illustrating where the force is and how strong it is.

So a charged object not only possesses charge and mass, it also emits an electric field. You can't see this field, this electric halo, but it's there and it has the potential to influence any additional charged object in its vicinity. Fields—invisible, secret, quiet, angelic ribbons of electric and magnetic force—are the heart of the power business.

It was the failure to know precisely where the fields were going that had impeded the design of better electric generators in the years after Faraday's discovery of induction.[20] In the 1860s and 1870s two big boosts came along to change this: first, a full mathematical theory of electromagnetic fields as derived by the Scottish physicist James Clerk Maxwell; second, the desire to market a system of electric lighting.[21] Maxwell's equations provided the mental push, the lighting companies provided the engineering push.

Out of this combination of concepts and cash came a powerful synergy. Better generators made for better lamps, which in turn led to still better generators. That's when Edison came into the picture.

He divided light. His system made possible *small,* household-ready bulbs. His grid energized mainly lamps, but this alone would not have changed the way people lived. The next big step forward in the ascent of electricity would be the use of better motors. If Edison had been the vicar of volts, Nikola Tesla was about to become the maestro of motors.

FORCE FIELDS

Sending electricity from Niagara to Brooklyn using direct currents, Tesla would argue, is an impossibility. DC power can be generated at low or high voltage, but because the voltage can't be altered once the electricity is on its way, DC current is usually operated at low voltage (around 110 volts), the voltage best suited to household lamps. But in this form the electricity would peter out long before it reached the Hudson River. This was DC's great shortcoming.

The power you send down the wire, the amount of energy per second, is the voltage times the current, essentially force multiplied by flow. You can make the current big or the voltage big. But keep this in mind: Electricity moving down a wire will always waste a bit of itself in heating up that wire. The wire resists the flow. The wastage of energy is proportional to the square of the current. More current, more waste. If, however, you make the current small and the voltage high, things would be different. The waste could be reduced to a manageable level. High voltage is what you need; low flow, high force. AC can be created at high voltage, transformed to even higher voltage, and then shipped without fear of devastating wastage. With AC power you can send electricity hundreds of miles. And at the end of the line, where the consumer and his bulbs are waiting, the voltage could be transformed back to the lower level for home consumption.

In the early days of the grid, DC power was not able to undergo these transformations. (The explanation of how transformers work is being saved for a later chapter.) Therefore, DC electricity could not be dispatched more than a mile or so because of the energy losses and because the voltage drop across the wire was proportional to the distance and to the current.

Mr. Edison, at this point, would object. High voltage is dangerous, he would assert. Besides, there is no worthwhile AC generator. And AC motors? Not worth a damn. It's not high voltage by itself that stops your heart, Tesla would have replied, but the current flowing through your body. As for an effective AC generator, I have designed one.

A generator, the machine that makes electricity in the first place (or, more properly speaking, that pumps electrical energy into wires), is mainly a motor in reverse. In a motor, currents flowing through competing coils of wire force a shaft to rotate. In a generator it's the other way around: A rotating shaft (spun up by a steam-fed turbine, say) forces currents to flow in the coil, in the manner of Faraday's flicker.

Tesla had indeed designed a better AC generator, but could he design a better AC motor as well? Early AC motors were poor. They stalled out, whereas DC motors were already being put to use in moving trolley cars through city streets, replacing carriages pulled by horses—along with their feed, stables, and droppings—with quiet electric transportation. Could Tesla do anything comparable to that? Could he succeed where others had failed? The allure of AC was tangible—being able to send power miles and miles was extremely attractive—but the motor issue was worrying engineers and investors. Furthermore, utilities had invested millions of dollars in wiring and equipment for DC electricity and wouldn't eagerly switch to another system. How could such a new thing as an AC grid come into existence in the face of such daunting opposition?

Tesla persisted. His first attempt to start an electrical company had failed, partly because he had been careless with patents, but now things would be different. He erected a lab, in which he built actual touchable prototypes of the visionary machines in his mind. When he turned them on, the motors made of iron plate and copper wire acted just like the motors he'd previously made in dreams. Just as Albert Einstein supposedly gained insight about the nature of space and movement by imagining what it would be like to run along next to a beam of light, so Tesla could, in that fever-pitched imagination of his, run along next to currents as they circulated around his imaginary machine. And what happens to those currents depends on the prevailing electric and magnetic forces. Swirling around inside the motor, like bottled energy, these two types of force feed off each other. When the current flows, it's nothing but excitement. A changing electric force can excite, or "induce," a magnetic force. Likewise, a changing magnetic force can induce an electric force. Tesla's motor would use this induction principle in a new way.

Tesla was a popularizer. "I much wish to tell you . . . I may say I actually burn with desire of telling you . . . what electricity is," was the way he once began a public address.[22] But this time his enthusiasm would not be allowed to overpower practicality. This time he would be careful with the patents. He would do things in a business-like manner. Neutral observers visited the lab, witnessed the performance, and could affirm the practicality. Now Tesla would tell the world.

He didn't seem to court fame, and yet he grew to enjoy giving public lectures on his favorite subject, providing spectacular demonstrations involving crackling sparks and bright lighting effects. At the podium he was patient and laid out a methodical promotion of his electrical ideas. Having a good presentational manner was useful since many of the early arguments over electricity were conducted at lectures attended by engineers wearing formal business attire.

Tesla's big chance came on May 16, 1888, when he spoke before the American Institute of Electrical Engineers in New York. It was then that he revealed his secret weapon, the induction motor, in which the current didn't just slosh back and forth in a single circuit, rising and falling with a sinusoidal shape, the shape of a moving water wave. Electricity in his plan would flow through several circuits. The currents would be slightly staggered so that, as in an automobile engine where various pistons are working in different parts of a stroke cycle (one piston going down just as another is coming up), the electrical motor would gain enormous efficiency and power. The electricity going into and out of the motor would flow in several circuits out of phase with the others but timed so as to be complementary.[23]

Inside the motor, if only you could see it, the magnetic force was swirling around like a tornado. Electric contended with magnetic, and magnetic contended with magnetic. It was a civil war between poles, north versus north and south versus south. One magnet would push off against another in a *yin-yang* of reciprocal, countervailing force. This microstorm was for the most part controlled and beneficent and was put to good use. This was Tesla's "polyphase" system, the centerpiece of his whole thought pattern.

Most of the engineers who attended Tesla's talk were—there is no other word for it—electrified. Tesla had his ideas, he had potent patents to hold off smart inventors panting to build their own AC motors, and he had the admiration of the profession, with the exception perhaps of the Edison DC faction and a few other also-rans searching for their own AC motor designs. What Tesla didn't have was money, at least not the kind of big money and influence needed to carry his venture forward into the arena that mattered, the commercial marketplace. To change the world he needed help.

Then, as if by fairytale connivance, the one man on earth who *could* help did turn up in Tesla's life. This man had already embraced AC electricity, had started to build AC grids, could see the merit of Tesla's claims, and had now come to buy Tesla's patents. There are some disputes even now as to how much was being offered, but between instant cash, stock, installment payments, and royalties to accrue from the sale of future motors, Tesla and his lawyers stood to take in more than a million dollars.[24]

DIVIDING VOLTS

Prefatory to anything else, Mr. George Westinghouse of Pittsburgh liked to examine the facts and weigh the alternatives before deciding on a course of action. His immense mustache, husky build, and riveting eyes would certainly give onlookers the impression of solidity. Although only 42 years old when he first encountered Tesla, Westinghouse was one of the wealthiest and most powerful businessmen in America. He had reached this status through a combination of shrewd financial calculations, benevolent management of associates, and bold investment in new technology. In a life full of eventful turns, the purchase of the Tesla patents was probably the most pivotal thing Westinghouse would ever do.

Like Edison and Tesla, Westinghouse had been a brilliant inventor, even in his 20s, and took great delight in the thought that he could produce a device that would make money *and* be a benefit to society. Like Tesla, he had seen some of his inventions snatched away and exploited by others, and consequently he learned to protect his work with protective castle walls of impregnable patents. Like the wizard of Menlo Park, the wizard of Pittsburgh attracted a devoted band of engineers who helped contrive not just workable machines but an industrial *context* for the machines.

Westinghouse, unlike Edison, did not cultivate a public persona. His manner of dress was not discussed, and his choice of breakfast food was not legendary (Edison was said to eat apple pie daily). Westinghouse had a memorable mustache. Yes, he was a blunt speaker, but he gave out no Mark Twain aphorisms. He did not leave behind writings, things like speeches or letters. His inventions and his grid would be his legacy.

Westinghouse had made his name by inventing something practical, a system for slowing the cars of a train using compressed air. This air brake saved lives and made money. It gave Westinghouse the cushion he needed to proceed with the founding of other companies based on new technology. In the early 1880s, while Edison was inventing the grid in New York and Tesla was dreaming of polyphase motors in Budapest, the man in Pittsburgh was barely dabbling in electrical affairs. Westinghouse naturally recognized the value of electricity but had viewed things chiefly from a railroad man's perspective. Electricity was something for signaling down the line or for setting a switch.

Then something came to his attention that changed his mind and altered his career. It was again the business of transmission. Unless they had a big voltage behind them, electric currents could not be sent very far. And even supposing the current started out at high voltage, how could it be brought back down to low voltage for home use? How could the electricity be converted from wholesale back to retail? What now caught the piercing eye of George Westinghouse was apparently a report of a device invented in Europe for converting electricity from one voltage to another.[25] Separate generators would not be needed for factories and homes. Everything could be accomplished by one generator and several of these converter devices, the modern name for which is *transformer.*

This versatile metamorphosis of volts could be carried out with AC but not DC electricity. Here was Westinghouse's chance. He would build his own grids with his own type of electricity. He would do with AC what Edison had done with DC. Edison had built an empire and so would Westinghouse. He too was an able negotiator, planner, marshaller of matériel. Edison had invaded Manhattan, stuck his banner in the earth, and claimed New York for his own. Since then he had seized other cities, but there were still plenty of opportunities. Edison, because his voltage couldn't go higher or lower, was earthbound. Although he had divided the light, he couldn't divide volts. Westinghouse could.

Westinghouse acted quickly. He sent an emissary to Europe to investigate the wondrous transformer. He obtained the rights, brought a model back to Pittsburgh, tried it out, stripped it down, adjusted the workings, and built it back up. A month later he incorporated the Westinghouse Electric Company. Two months after that, in March 1886, he unveiled the first centralized AC grid in America, in Great Barrington, Massachusetts.[26] The press coverage for this event was not as intense as it had been for the Pearl Street debut, but at least Westinghouse had made his point. Eighteen months later AC grids were operating or under construction in dozens of cities. Thus was launched the war between the currents, AC versus DC.

No matter what Edison said, the new grids for AC electricity kept coming. And why not? People previously beyond the reach of the grid were now happy to receive the benefit of electrification. But what about those shifting currents, always flipping left and then right? Doesn't that mess things up? Well, even though AC power reversed itself many times every second, the human eye could not see any flicker in a lightbulb. In most respects a customer would not know the difference between electricity that was steady-on and its rival which, every 60th of a second, would diminish in size, stop, reverse, and then build up again.

One great problem remained for AC grids. The energy-by-wire electrical revolution had begun as a way of lighting lamps, but it had promised also to power all those machines that had formerly required steam—lathes, drills, stamps, mills, compressors, and so forth. DC motors were starting to fill that promise. What about AC? Where was *its* motor? This is why Westinghouse so avidly came to his historic compact with Nikola Tesla who, just at this precise moment, promised an efficient motor that thrived on AC electricity. Westinghouse not only bought the motor, but he also bought Tesla himself.

Tesla proceeded to Pittsburgh to work with Westinghouse's own engineers to integrate the motor and its associated dynamo into the existing AC framework. It was then that the true startling nature of Tesla's designs became evident. First, the standard AC grid of the day switched back and forth 133 times per second. Tesla argued that this was too fast for his motor, which operated much better at a rate of 60 cycles per second. Second, the standard AC grid used a single pair of wires for hooking up machines, just like their DC counterparts. But Tesla's motor required two or even three circuits and therefore needed extra wires. Would the grid have to be rewired just because Tesla said so? The engineers balked. They naturally wanted to make everything work with just the one circuit. Tesla, whose creativity was better suited to a solitary environment, left Pittsburgh and retreated back to his base in New York. Disenchantment with his motor system was starting to grow. Had Westinghouse been wrong to trust Tesla?

The stalemate did not happen at a convenient time. The competition for grid business was fierce. Furthermore, a prestigious contest had been announced. The prize was a contract for wiring a world's fair to be constructed in Chicago to honor the 400th anniversary of Columbus's voyage to the new world. The Columbian Exposition promised great publicity to any company that could deliver electricity for the most magnificent gathering of international wares, inventions, and folkways ever assembled in one place.

There was at this time a more ominous development. The American economy, like any other complex human cultural institution, has its ups and downs. The stock market crash of 1929 and the ensuing depression of the 1930s together are perhaps the most famous business downturn of the 20th century, but there had been earlier depressions. The year 1907 saw a bad economic slump, and the early 1890s were another such period.

So it was that Edison's empire, still strong in terms of city grids and sales of manufactured goods, was poor in ready capital. As an inventor, Thomas Edison was without peer. As a showman and cheerleader for electricity and as an inspirer of men and orchestrator of great events, he was remarkable. As a businessman, however, he was only fair, and this, coupled with the straitened economy, was causing problems. Edison might still be a Napoleonic figure, but if so he had entered the Waterloo phase of his electrical career. Still an imposing personality, he was gradually being overtaken by a predominating array of forces standing against him. He had become, even with all those creative ideas springing from his head like lightning bolts, a bit in the way.

In 1892 a triumvirate ruled the American electrical market: Edison, Westinghouse, and the firm of Houston-Thomson, another great pioneering electrical conglomerate. It was expected by many that the first of these companies would buy out and merge with the third, with Edison in command. Things turned out differently. Partly through the financial maneuvering of J. P. Morgan, the two companies did merge but with the Houston-Thomson management team in the superior position. Edison was to be paid handsomely but would lose control of the company he had founded. The reformed entity would not even bear his name. The newly crowned king of electrical manufacturers was (and is still) called General Electric, or GE for short.

Westinghouse was also being pressed by investors. Successful in gaining customers but troubled by a weak economy and underperforming investments, he was forced by events into an unpalatable expediency. While Edison had seen his name effaced from the company shingle, Westinghouse's ordeal, his personal mortification, was to have to go to Tesla and ask him face to face whether he, Tesla, would forego his potentially lucrative royalties on his so-promising motor design. Only in that way, Westinghouse argued, could the company continue. Tesla was profoundly appreciative that Westinghouse had believed in his engineering concepts and was turning what had been in Budapest only a vision into a practical reality. Tesla magnanimously renounced his claims. In what must surely be one of the greatest sacrifices in the history of high-tech patent rights, he gave Westinghouse a reprieve.[27] Tesla received a lump settlement in lieu of future royalties.

In the 1890s slump, Edison lost his company. Through Tesla's generous act, Westinghouse was able to retain his. This left him free to do battle with General Electric for the right to wire the world's fair in Chicago. Beyond this magnificent enterprise loomed an even larger project.

THE AGE OF DISCOVERY

In 1492 Christopher Columbus hove up on the sands of the outlying islands of the Americas. From the European perspective, this was a vast discovery. From the native perspective, Columbus's arrival was an invasion and a pestilence. What the Europeans found was a landmass no less than two continents in extent, a variety of previously unknown peoples, and enough new botanical and zoological species to overflow a shelf of books. What the Indians got was mostly disease and subjugation. For better and for worse, the Spanish expeditions were staggeringly important for world history.

Four hundred years later the anniversary of this momentous encounter was being celebrated at the greatest world's fair ever held. The Columbian Exposition of 1893 honored the Genoan admiral of the high seas, but it also honored the rapid rise of the fair's host, Chicago, a city that had become a leading metropolis in the land. And befitting such a great city, three other major lakeside cultural institutions were coming into existence just then: the Chicago Symphony Orchestra, the Art Institute, and the Field Museum of Natural History.[28] When cultural historians look back at the fair, however, what they see even more clearly than the legacy of Columbus or the upstart growth of Chicago was the glittering arrival of electricity.

To see why, just compare the last great world's fair to be devoted to steam power rather than electricity, the one held in Philadelphia in 1876, with the one in Chicago in 1893. The Philadelphia exposition would virtually shut its doors at sunset each day, whereas in Chicago the magic *began* at sunset because that's when the power of electricity revealed itself in full force.[29] More than 27 million people attended the fair, and the most splendid sight, the *creme de la creme,* was the electrical pavilion. On display was the world's largest engine and the largest assemblage of powered machines anywhere. Keeping everything energized, including 100,000 bulbs, was the world's largest generating system.

The extravagance of both the quadricentennial observance and the accompanying light show called forth some grand comparisons. The intrepid men of the age of discovery, the first Europeans anyway (trappers, soldiers, missionaries), had moved up uncharted rivers, across dry plains and thick forests. The explorers of electricity, centuries later, were now ushering in their own age of discovery beneath city streets and in the recesses of those magnificent spinning, shaking, thundering machines. These engineers were colonizing not prairies but lighting districts: first Pearl Street, then other city centers, later still the outlying regions, maybe someday other worlds. They were discovering not gold nuggets but something more valuable—superior current-carrying materials. If Ponce de Léon had sought the youth-preserving elixir of El Dorado, well, then, on the electrical side it was said that voltage rays were sure to have healthful effects of their own.

There had been a spirited battle to win the right to light the fair. Not only did the contract represent an enormous business proposition, it would be a public relations triumph for the company illuminating the new El Dorado. Westinghouse won that contract, mostly by entering a money-losing bid. He hoped to recoup losses in the form of enhanced prestige and future contracts for grids and electrical appliances. Westinghouse Electric Company, incorporated only six years before, proceeded to build one of the technological wonders of the world in Chicago. A 2000-horsepower engine, the most powerful AC device in the world, joined by a dozen 1,000-horsepower engines, supplied an awesome blizzard of electrical current. With more than 10 times the number of bulbs of the Paris exhibition of 1889, the Columbian Exposition of 1893 consumed three times as much power as the surrounding city of Chicago itself.[30] Westinghouse's display at the fair was of course quite conspicuous, but even GE (which couldn't afford to sit out an occasion like this) had a large exhibit space of its own, where it showed off the world's largest lightbulb, 8 feet tall, and Thomas Edison's "kinetograph," which "transmitted scenes to the eye as well as sounds to the ear."[31]

At the exposition the electrical building was 690 feet long, and its displays were described as the most novel and brilliant sight on the grounds. Here is what the guide to the fair had to say:

> Of all the separate World's Fair departments, the Electrical has a peculiar novelty and freshness in the popular mind. It differs also in one supreme particular from all of the others. The rapidity of electrical development finds no parallel in any other range of discovery. To the electrician ten years is a century, and even in one year all of his pet theories may vanish under the light of some new discovery. Further, the science of electrical development has advanced just far enough to teach the electricians that they are merely on the threshold of unbounded worlds of knowledge. The present exhibit, marvelous as it stands when compared with electrical knowledge ten or twenty years ago, may prove to have been crude and insignificant before the rounding out of the present century.[32]

What Westinghouse had lost on his operations at the Columbian Exposition he was shortly to regain in stature and business experience. For another thing, the electrical setup at the fair had been mostly of the AC type. The war between the currents had basically been settled. Edison had lost not because his electricity was direct current but because it was low voltage. It wasn't adaptable for lots of applications because it couldn't travel far and it wasn't divisible.

DC was not exactly going away overnight, but AC was now champion. Furthermore, the Tesla conundrum, whether to incorporate the greater complexity necessitated by his multicircuit (polyphase) designs, was settled in Tesla's favor. Even his lower operating frequency of 60 cycles per second (also called 60 hertz, or just 60 Hz) was adopted (although Europe would settle on a convention of 50 cycles per second). Tesla-style dynamos and motors, even with the extra wiring, were suddenly in demand with customers. Ironically, GE, the company founded by Edison, grudgingly began to market AC equipment of its own. How the world had changed in just a few years!

THE ELECTRIFICATION OF HENRY ADAMS

Perhaps things had changed too much. Henry Adams thought so. The two most salient facts about Adams are that he was a descendent of two U.S. presidents and he was a historian who wrote fussy, self-conscious, worried, but very perceptive prose.

Adams loved to write about change, and that is why he went to the Columbian Exposition. During his career he wrote three great books about change. One was about the impact of the Gothic age on Medieval France in the 12th and 13th centuries. Another was about the impact of government policy on the young U.S. republic at the beginning of the 19th century. The third book, the one that concerns us here, an autobiography called *The Education of Henry Adams,* was about the impact on Adams himself of the social and technological changes at the end of the 19th century.

Not the least of those changes was the coming of electricity. In fact, Adams's principal symbol for the age was the dynamo powering all those clanking, clamorous machines. Adams liked to contrast this emblem of modernity with the emblem of Medieval spirit, namely the Virgin Mary and the many Gothic cathedrals dedicated to her name.

Like the millions who had come to the Chicago fair, Adams was both delighted and concerned. Unlike the millions of other attendees, he could render his ambivalence in the form of insight-filled philosophical perspective. Here he is, writing about his bafflement and sense of dislocation as he entered the Electrical Hall:

> Some millions of other people felt the same helplessness, but few of them were seeking education, and to them helplessness seemed natural and normal, for they had grown up in the habit of thinking a steam engine or a dynamo as natural as the sun, and expected to understand one as little as the other. . . .

Adams wanted to understand. He is one of several conscionable characters in this book who asked the question, "This looks good, this electrical apparatus, but are we sure we're doing the right thing?" It must be said that he did not reject outright the electrical behemoths. In fact, he was rather fascinated by them and wanted to know more about them. For him the pace of technological advance was getting to be too great, a feeling that we, in our time, know well. He was the kind of person who prided himself on keeping up with developments, but gradually he developed a feeling of falling hopelessly behind.

Adams's world of 1893 had not yet experienced aviation, or radio programming, or antibiotics, or the Internet, but he could sense that the new technology would redraw boundaries, change rules, and otherwise alter the conditions of settled existence.

> Men who knew nothing whatever, who had never run a steam engine, the simplest of forces, who had never put their hands on a lever, had never touched an electric battery, never talked through a telephone, and had not the shadow of a notion what amount of force was meant by a watt or an ampere or an erg or any other term of measurement introduced within a hundred years, had no choice but to sit down on the steps and brood. . . .[33]

GRID HONEYMOON

The electrified future, foretold and glorified on the shores of Lake Michigan in the artificial environment of the Chicago exposition, would now begin to take practical shape some 800 miles away, not far from the shore of Lake Ontario. It was at Niagara Falls that the Almighty had providentially caused the collective waters of the Great Lakes to undergo a mighty descent on their way toward the Atlantic. It would be at Niagara Falls that electrical technology would take its biggest step forward since Pearl Street.

Niagara represented the biggest water gradient in the eastern United States, and it was evident that a worthwhile share of the energy in all that falling water could be converted to usable electrical energy. The officials in charge wanted to use the very latest engineering concepts in making an immense set of turbines, unprecedented in size and power, for creating more electricity in one spot than had ever before been done. And then they wanted to send that energy to faraway places. Recent tests, including one at an electrical fair in Frankfurt, Germany, had shown that electricity could travel more than 100 miles.[34] If the voltage were high enough, there appeared to be no limit to how far the grid could stretch.

The electrical setup at the Columbian Exposition had been make-believe: electricity energized an immense light show and a Ferris wheel and electric boats wafting across an excavated pond. This had been a fairyland grid. Now the competition, again pitting mainly Westinghouse against General Electric, would be for building a much grander grid, a grid that would last, a grid that would change the nature of heavy industry. Even people in New York City, at the other end of the state, were intrigued by what was about to happen in Niagara Falls.

Don't worry, the beauty of the falls would not be marred by the presence of the huge apparatus. Everything would be out of sight. The needed water would be discretely diverted from a place upriver, leaving plenty still for the newlyweds on their honeymoons to look at but enough to empower a giant hydroelectric plant. (At night and in the off-season, the power company was permitted to take twice as much water as usual.) The ambitious plan, when fully implemented, would include water falling through tunnels leading to 20 dynamos, each larger than any previously in existence.

The launch of Edison's Pearl Street plant and the electrical concession at the Columbian Exposition had each needed roughly a $500,000 investment, but the Niagara contract would come in at around the $6 million level. Westinghouse and GE had sensibly decided on a truce as far as patent lawsuits went, but they still fought hard over the Niagara deal. In the end Westinghouse again triumphed over its great rival, at least in the matter of the powerhouse, but GE was not left out. Its share of the Niagara prize was the job of transmitting the power to distant customers.

You will notice that I use the word *grid* a lot in this book. Depending on the context, it will sometimes refer to the wires or electricity making capacity of a particular utility and sometimes to the totality of the electrical activity in a whole region, or even (at a later stage) the whole nation or world. The notion of an aggregate grid is pertinent to the hydroelectric project at Niagara Falls since the amount of power produced (or about to be produced) was so huge as to vastly exceed the needs of the immediate locality. It would be advantageous to hitch the Niagara grid to other grids, even as far south as Manhattan, in order to make use of all the hydropower.

Thus the wider grid, in the sense of being a network of networks, was about to enlarge itself in a spectacular way. Edison had built a mile-wide grid using direct current. Then Westinghouse had pioneered the simple alternating-current grid which, because of its high-voltage oomph, could stretch out for tens of miles.

Tesla's contribution was not to enlarge the grid in its physical extent but to enhance its value through better use of electric motors. On the day they threw the switch at Niagara Falls, it was Tesla's polyphase wiring they were using, Tesla's frequency (60 Hz) they were using, and Tesla's patents immutably inscribed on metal plates that were bolted to the side of the dynamos—in effect, the license plates for the new grid.

Since Pearl Street, the spread of electricity had been impressive but still limited in scope to the upper strata of society. Consider this snapshot. Take a look at where electricity went in Berlin around this time: the top customers, in descending order, were theaters, banks, restaurants, shops, hotels, and streetlights. At the bottom were industry (2 percent) and homes (1 percent).[35] After Niagara Falls this ranking would change in a big way. The Niagara grid was about to do more than light a lot of lamps and grind a lot of wheat. Plentiful Niagara electricity would soon turn the Buffalo environs in upstate New York into the greatest high-tech corridor of its day. It would be the Silicon Valley of the 1890s and the electrochemical capital of the world.

With cheap power coming from the falls you could give tin spoons a sparkling electroplating of silver. To extract aluminum from its ore, it would be better to use finesse; instead of brute pulverizing and melting, the smart way is to tickle the ore with an electrical current, causing the surrounding avaricious silicate to relax its grip on the metal atoms. A production rate once measured in ounces would soon be measured in tons or thousands or millions of tons.

On November 16, 1896, the energy vested in water coming from as far away as Green Bay, Wisconsin, water eager to return to its ancestral ocean home, having worked its way through the Great Lakes to the west, took a shortcut through the apparatus of the Niagara Cataract Company. The bottled lightning made from that water was rapidly transformed up to 20,000 volts, the better to race unimpeded southward where, having dutifully remained within its wire and sprinted the distance in one ten-thousandth of a second, it was converted back to sensible voltage in preparation for making its appointed rounds. The end of the beginning of the grid was about to take place. In Buffalo that morning, with thousands of curious onlookers ready to be impressed, at just the right time and powered by a potency 20 miles away, the horseless trolleys were impelled along their tracks.

3

MOST ELECTRIFIED CITY

The setting was Delmonico's, the same restaurant that had catered Edison's feast announcing the launch of his grid. The ceremonial occasion now marked not an arrival but a departure. Several big ironies hung in the air. First, the man being honored was not at the threshold of retirement. Instead, he was at the early stage of what would be a long career. Second, although this was New York City, birthplace of the Pearl Street powerhouse and effectively the headquarters for world electricity, the young man was about to move away. And he would take the grid with him.

Few would have recognized this event as a passing of the guard. Edison had invented the grid, invented the *system* of wire, socket, plug-in bulbs, feeder cables, and the dynamos to power it all. Actually what had powered everything was Edison's own personality. He had maneuvered among politicians, bankers, and engineers to create the massive, flexible organism we now call the power network. Westinghouse then reinvented the grid by resorting to pulsing alternating currents and transformers, those hulking machines that jack voltages up and down. Tesla then rewired everything all over again with his multiphase circuits and his motors that harbored internal blizzards of magnetic force. All three of these men were alive and well. The best of their electrical work, however, was behind them.

And the new man? What did he do to deserve a testimonial dinner? Strangely, the evening would be significant not because of what the upstart had done already, which was considerable, but for what he said he was going to do next.

PARTING SHOT

This book is short, but the grid is long. Blithely bypassing dozens of other names who in a longer account would have merited a place in a larger history of the grid, the narrative has so far concentrated on three founding fathers: Edison, Westinghouse, and Tesla. To this short list we now add a fourth. Actually, he's already been on stage. He played a significant part in the saga of volts, but you wouldn't have noticed. He was there when the first electricity raced out of Pearl Street. Although only 22 at the time, he was indispensable to Edison and Edison's plans. Now he had plans of his own.

Not well known at the time by press or public, Samuel Insull made all the parts of the Edison machine work in synchrony. Having arrived in New York from Britain only shortly before Pearl Street, he was hired to be Edison's personal secretary, but his work quickly expanded to include organizing the office, overseeing finance, arranging loans, running errands, and going on inspection tours with his boss to lathe works at three o'clock in the morning.

In the wake of the Pearl Street debut, business had boomed and Edison directed Insull to consolidate and move the various manufacturing ventures, many of them scattered around New York City, up the Hudson River to Schenectady. The Edison vision of universal electricity had adroitly been brought into reality, but the sudden call for electrical service and electrical apparatus had overwhelmed the Edison company. It simply could not keep up with demand; that is, until Insull created a small revolution in the mass production of parts (long before Henry Ford was to be credited with the concept) that allowed the volume of shipped products to go way up and the price per part to go way down.[1] When the Edisonized central grids for cities all over the United

States—Boston Edison, Detroit Edison, and so forth—had their own Pearl Street start-ups, many of the parts came from Insull's factory.

There seemed no limit to Insull's ingenuity, energy, or resourcefulness. After making a great success of the Schenectady operation, he was stuck in the middle of the incorporation melodrama surrounding the creation of General Electric. He had foreseen the need to consolidate companies but had also been loyal to Edison and was sorry to observe his mentor eased out of control and the "Edison" removed from the name of the new company.

Things had turned sour. Indeed, after serving at Edison's side for a dozen years and coming to like the bustle of New York City, Insull resolved to leave town. The instigators of the corporate makeover had recognized Insull's skills and designated him as the prospective first vice president—in effect the number three man at GE—but he preferred to be the number one man at another company, even if this meant receiving only one-third the salary. And he was young: Having already had the equivalent of several full-length careers, he was only 32 years old.

At the Insull farewell banquet, all of his "most intimate friends and most intimate enemies" were present.[2] Who knows if it was the wine or an undercurrent of bitterness, but that night an uncharacteristic small cloud of unguarded speech was to pass Insull's lips. His new employer, a utility company called Chicago Edison, was a dust mote compared to GE, and yet Insull brashly boasted that he and his new company would someday overtake GE.[3] He would overtake everyone. Many in the room would have laughed uneasily at this awkward moment because they knew of Insull's steely resolve and his plentiful ability. Certainly they weren't laughing a few decades later when this young immigrant from Britain had indeed built his own electrical empire and was one of the most powerful men in America.

Insull does not now command much attention in the history books, which is a pity. For consider, he probably had more than any other individual to do with the way your home is wired, with the variety of appliances you own, with the nature of the electrical bill you receive each month, with your morning commute on mass transit, even with the process by which your power is produced and distributed, financed and regulated. And then, at the height of American power, he suffered a fall of epic proportions.

TURBOCHARGED GRID

Insull grew up in London. When he came to New York he felt that his adopted city was raw by comparison and too full of energy. Later he changed his mind and came to embrace the vitality of New York, a sentiment many share to this day. When next he moved farther into the continent, to Chicago, Insull encountered even more rawness and more energy. Here, stuffed with stockyards, railroad sidings, granaries, and smoking factories, was the fastest-growing city in America. Dismayed all over again by the barbarian coarseness, Insull shrewdly inoculated himself against taking fright and flight by signing a multiyear contract with his new employer. As a conspicuous act of good faith, he invested a lot of his own money in company stock. He would force himself to make a go of it in Chicago. No matter what happened, this was now his home.

When Insull arrived, Chicago Edison had a few thousand customers, in a city with a population of a million. This wasn't enough for a really ambitious man. He needed more power, in several senses of the word: more power from bigger generators, more power over his retail environment, more political power. His first natural advantage—his ante into the great electrical game—was the exclusive rights to the strategic patents on Edison merchandise such as bulbs and switching equipment. With these Insull started to elbow his competitors out of the way. If another utility couldn't buy the machines and parts it needed because of patent limitations, it wasn't going to have much of a chance. Insull had started by competing with his rivals, but his preferred method was to buy them out, one by one. Like the fictional smart-aleck Charles Foster Kane outdoing his newspaper rivals by buying up their best writers, Insull bought himself power stations, licenses, and customer contracts until his was the only grid left.

Chicago Edison's earliest central station—its equivalent of Pearl Street—was the Adams Street plant, built in 1887. Insull's first order of business after settling into the job was to build something bigger, in fact the biggest power plant in the world. Size bought you efficiency, and more efficient generation lowered the unit price of electricity and this brought in more customers. More demand necessitated more generators. In this way, even the colossal new plant quickly became fully subscribed. The mighty piston-driven steam machines already filled every available berth in the building. These electrical leviathans heartily shook the ground, pushing the parameters of mechanical and electrical engineering to the utmost. Insull needed a new masterpiece of engineering, and this would cost money.

Even in the middle of a national economic slump, Chicago was humming. Insull had made himself a master of money as well as a master of machines. He succeeded in the loan market, often by using his extensive connections in Europe. Six years after Insull had taken over, Chicago Edison's load, the aggregate amount of electrical gadgets receiving current, had climbed by a factor of 10. For Insull this was still too small. The load was 10-fold, yes, but the number of customers had only doubled.

Insull had the monopoly he wanted, but he still acted as if the competitors were at his heels, and in a way they were. Many homes were still getting along with gas lighting. Electricity was something fancy hotels and restaurants had but not ordinary homes. Electricity was desirable, fashionable, modern, but it was not yet a necessity. For many citizens electricity was still too expensive, and this bothered Insull.

In his hands the existing reciprocating steam engine design had reached its size limit, and something new would have to be found. The name for the new thing was turbogeneration. In the old engines, the horizontal motion of a piston was converted through linkages into the rotary motion of the generator shaft. Picture a giant steam locomotive that doesn't go anywhere. A turbine is different. There the roiling steam presses against blades mounted on the shaft itself. There would be no back-and-forth motion, only the rotary movement of the shaft. A turbine is a sort of windmill in which the force comes not from wind but from steam.

Was Insull the first to use turbine generators? No. Turbines had been used on a small scale—to propel ships and power small grids—but Insull was going to be the first to use turbo on a big scale. He didn't just want turbines but unprecedented turbines. General Electric, which was in the business of eagerly supplying utilities like Chicago Edison, had to be coaxed into filling Insull's request because it seemed like a risky venture at the time.[4] The result, the world's largest generator, cranked out 5 million watts, 5 megawatts of power. It was emplaced in its own palace of power, the newest, most grandiose, generating station in the world. On the big day, when it came time to throw yet one more of those switches that mark the growth of the grid, Insull's chief engineer told his boss to stand back, in case there was an explosion.

"There is just as much reason for you leaving here as my leaving here," Insull said.

To which the engineer replied, "No. My being here is in the line of my duty."

Insull, who had not only pressed the mechanical designers to a feat of extreme engineering but also rallied his investors and board of directors nearly to the breaking point, had the last word. "Well, I am going to stay. If you are to be blown up, I would prefer to be blown up with you as, if the turbine should fail, I should be blown up anyway."[5]

There was no explosion. The machine occupied a tenth the space and weighed a tenth as much as its predecessor. It not only worked, but very quickly was generating 10,000 horsepower, double the power produced by any steam-driven generator ever built.

ENERGY MULTIPLICATION

What is 10,000 horsepower? How impressed should we be by Insull's new turbine? It's no use quoting a number like that without giving you a sense of what it means. Already at this point in the narrative, the generators have gotten pretty big, and they're going to grow even bigger. Actually, 10,000 horsepower is already too big. We should start small. In fact, let's start with a fraction of one horsepower.

In examining the dynamics of food production and energy use in Colonial America, sociologist David Nye begins with this simple proposition: A man grinding wheat is exerting himself (spending energy) at a rate of about one-tenth horsepower. In other words, a draft horse pulling a shaft that rotates a millstone can do the work of about

10 men.[6] Ever since Adam and Eve first earned food by the sweat of their brows, the struggle to acquire sustenance has been just about the most important daily imperative for humans, as indeed it is for the rest of the animal kingdom. Creatures in the wild will always be hunter-gatherers, whereas humans at least were able to employ their cerebral circuitry to develop, first, agricultural methods and, later, technological implements to enhance the available muscle power.

Nye gives several examples of multiplication of effort through mechanical contrivances. Wind power: in the year 1500, the sails used to propel a 100-ton ship would produce perhaps 500 horsepower, a factor of 50 times greater than the collective muscle power of the human crew of that ship.[7] Water power: a typical river-driven mill in colonial America was about 4 horsepower. Is it any wonder that often a mill was built before the town was built? The 4-horsepower mill could grind the same wheat that would take 40 men to produce. With a mill on the job, wheat could be turned into flour by one man (the mill operator) while the other 39 men could be freed up for other jobs. If the mill employed a turbine rather than a wheel, the power could be doubled.

So the great Energy Pharaoh, Samuel Insull, commanded with his 10,000-horsepower generator the equivalent of 10,000 draft horses pulling at the yoke or, even more stupendous for the mind's eye to fathom, the labor of 100,000 people toiling away on hand-cranked mills converting grain to flour.

Bigness begets bigness. Once the principle of large turbogenerators had been established, the old power limitations seemed to disappear. It was like the transition from propeller to jet aviation. Chicago and utilities around the world starting building bigger and bigger turbines. First, 5 megawatts, then 10, then 35. With this kind of capacity, increased demand could easily be met. For that reason the making of electricity will, for the moment, be of less interest to us than other more subtle aspects of the power business.

Insull, the engineering juggler, could easily convert between alternating and direct currents, between high and low voltage, and replace piston with turbine. All of the new hardware innovations contributed to a vast economy of scale: generally speaking, the larger and more flexible the machine, the more efficiently it would perform. With certain operations, and the grid is a good example, the overhead costs of doing business largely stay the same, so the more product that is sold or the more customers serviced, the more the cost of overhead can be amortized.

The price of electricity was coming down, but this by itself wouldn't be enough. Human nature had to be taken into account.

THE SOCIOLOGY OF THE GRID

Move from the Engineering Insull and the Management Insull to the Anthropological Insull: How did society use the grid?, he asked himself. How much energy was extracted from the grid, at what hour, and to what end? The metering scheme first introduced into the U.S. market by Chicago Edison was able to get readings over shorter time intervals, and this revealed the fascinating anatomy of power around the clock. Department store usage, for instance, was big during its open hours, between 8 a.m. and 6 p.m., and almost nil for the rest of the diurnal period. Garages where electric vehicles were charged up kept reciprocal hours; their operations stretched from 8 p.m. until 7 a.m. Manufacturing power needs were pretty well distributed across the workday, with a noticeable dip at noon when workers broke for lunch. For homes the peak was in the evening when families were actively together.

Electricity was still too expensive, and Insull could see why. Like a cardiologist gaining vital evidence about a patient by reading an EKG scan, Insull could interpret a lot from his load curve. A generator working for only a few evening hours when homes wanted electricity for lighting was an underperforming capital investment, an underused piece of equipment. That machine's load factor, the fraction of time it was actually in use, was poor. Insull illustrated this point in a lecture by analyzing one particular housing block. This little cross-section of Chicago contained apartment customers and garage customers. Here is Insull speaking before the Young Men's Christian Association:

> If you take each customer by himself, that is, each apartment by itself, the use of energy in each separate apartment is so slight that the investment to take care of that particular customer, if you trace it back to the generating station where the power is produced, would not be used on average more than between six and seven per cent of the time. But so varied are the ideas

> of human beings, and they so seldom do the same thing at exactly the same moment that, if you take the whole 193 apartments together, and then find out how much energy as a whole they use at one particular moment, the fact is developed that the diversity of their demand is so great that instead of using your investment only between six and seven per cent of the time, they use your investment, when taken as a whole, twenty per cent of the time.[8]

And if you were to add in the stores around the block, you would find that the investment was being used at the 30 percent level, which is better than 6 percent but still not as good as 100 percent.

Using the investment fully and gainfully was vital not only to staying in business but to expanding. Moreover, with Insull efficiency seemed to have been not just a business imperative but also an intellectual pleasure and a moral attainment, as if by wringing out the last drop of electrical utility from a pound of coal one were fulfilling a quest begun when the coal was first fabricated in the earth eons ago.

So varied are the ideas of human beings. . . . Insull saw that he could use human diversity to help smooth out his load curve. The way to increase productivity was to diversify his customer base. To get customers to use electricity whenever possible in off-peak times, Insull offered rate incentives. All of this was part of his grand scheme to increase profits by *lowering* rates. Go after a larger market share, or in this case create a market in the first place, in the hope that increased sales volume would make up for the lesser charges per unit of production.

To make his economy-of-scale operation thrive, Insull needed to conquer more territory. He won contracts for powering the trolley business in Chicago. Formerly these electric trains had had their own power supplies, but Chicago Edison essentially made them an offer they could not refuse. With Insull's electricity providing the traction for pulling trolleys up and down long boulevards, all his expensive underused turbines suddenly had something to do during two formerly slack periods—the time just before the work day began, when people were commuting to their jobs, and the time just after the work day ended, when people went home.

These two segments in Chicago's daily rhythm, what we now call the morning and evening rush hours, greatly helped smooth out the grid's load curve. Trolleys, at least for a time, were the biggest users of electricity in the United States; they accounted for two-thirds of Insull's load.[9] And what had all the electricity replaced? Well, a few decades earlier the way workers got to their jobs was on foot or in omnibuses pulled by 100,000 horses nationwide, at an average speed of 5 miles per hour, and leaving behind on the streets a million pounds of manure every day.[10]

The price of electricity kept falling. Chicago Edison had built more efficient turbine generators, had brought the grid into many new neighborhoods, had acquired a familiarity with the hourly rhythms of the grid, and had lowered rates to induce customers to sign up. Yet they resisted. Electricity for the average citizen was still too expensive. Samuel Insull never gave up. In fact, he was just getting started.

FAUSTIAN BARGAIN

In 1898 Insull was elected president of the National Electric Light Association. Speaking at the association's convention, he proclaimed his evangelical message of lower rates as the way toward higher profits. He had been one of the greatest proponents of the necessity for monopoly and indeed had arrived at that condition in Chicago when his final competitor succumbed.[11] His argument? Utilities, to deliver energy at the lowest rates, had to be free from wasteful duplication of expensive wiring and heavy equipment. He was far ahead of his brother utility operators in recognizing, however, that operating as a monopoly brought with it the need for public regulation of privately owned utilities. Yes, he would say, you heard me right. The customer, and just as importantly potential investors, will have greater trust in the course we steer if our operations and our profit margin are negotiated in public view.

Commonwealth Edison, as Insull's company was known by 1907, had done as much as it could on the production side. Now what needed work was the demand side of the grid equation. Insull did not, and never could, have enough business. When he started in Chicago, he had 5,000 paying customers. By 1898 it was 10,000; by 1906, 50,000; by 1909, 100,000; by 1913, he had 200,000 ratepayers. Commonwealth Edison was larger than New York Edison, Brooklyn Edison, and Boston

Edison put together.[12] His company had become the model for utilities all over the country.

Trolleys were electrified; factories increasingly had powered equipment; and electric lights illuminated the streets, train stations, and the better hotels and restaurants. But for ordinary people, electricity was still special. Reaching into homes, getting people to change their habits, would take time.

Insull's first attempts at advertising the benefits of home electricity were aimed primarily at rich folks. He started a monthly magazine, *Electric City*, and a store specializing in electric appliances. "Buy something electric for Christmas" was the slogan one year. He touted a portable "Electrical Cottage," displaying various impressive wares. Then came the ploy that earned Commonwealth Edison a place in the history of the early golden days of mass advertising. It worked like this. A cart was sent through residential neighborhoods. The cart was piled high with electric irons, and the man in charge was offering a bargain that would rival the tasty apple offered to Eve in the Garden of Eden. Ladies, allow electricity into your home—installation charges spread out over two years, with no financing charges applied—and replace your old, cumbersome flatirons with modern electric irons. We have 10,000 new irons to give away.

Here at last was temptation difficult to resist. Electric toasters or electric fans—these things were luxuries, hallmarks of a pampered life. But ironing—well, everyone needed clean clothes. After lighting, the next most used electric contrivance in the home was to be the iron. Heating and maneuvering the old flatirons from stove to shirt, back to the stove for reheating, and then more shirts, was muscle-aching and finger-blistering work, especially in summer. With the electric iron, there was still plenty of work to be done, but the energy for heating the irons arrived in thin wires. There would be much less hefting and sweating.

In this way electricity gradually came into Chicago homes. It didn't happen all at once but in stages. The utility kept building more power plants, kept extending its lines into more neighborhoods, kept lowering its rates. Electrical current was no longer some novel, invasive vine but something more like a native species. Even if you didn't yet take the service, you knew people who did, and you were aware of the wires strung overhead from poles. You looked at them every day and wondered how exactly the system worked. You couldn't see the current flow and there wasn't any detectable smell as there was with gas. Now and then a low hum could be sensed, but otherwise there wasn't much of a sound.

What the utility was offering was highly attractive and increasingly affordable. Not that a housewife would be overly indulgent with her electricity. No, it would be doled out sparingly. Just as she would continue to use gaslight in the evening, saving the electric lights for when company was over, so the lady of the house would often go on using the flatirons in winter—you had the stove going anyway, so why not use the flatirons?—and save the electric irons for summer. Electricity in your home was a thing to be noticed and nourished. It was not taken for granted, at least not yet.

What was electricity like in those years? Here are some particulars. The average apartment with electricity had a dozen sockets. Houses had twice as many. The cord bringing power into an appliance had to be screwed into one of the dangling light sockets since the standard plug didn't come along until the late 1920s.[13] The bulbs used were typically rated at 50 watts but gave off only a fraction of the light emitted by a comparable bulb today.

We've looked at lamps and irons. Presently we will look at the third great domestic product of the electricity age. It took longer to develop than the bulb or the iron, but once it had been perfected it caught up quickly. Five years after it was introduced, 60 percent of Chicago homes had them. Five years after that they were everywhere. The new invention was practically more popular than the grid itself.

ON THE AIR

This book is about sending energy by wire. We can't yet send food or other bulk goods by wire, but it is extremely useful to transmit information this way. Here are six great electrical-information gridlike inventions: (1) *telegraph*, the movement of binary code via wires; (2) *telephone*, the movement of voiced sounds via wires; (3) *wireless*, the movement of telegraph-like binary code via electromagnetic waves sent through the air; (4) *radio*, the movement of voiced telephone-like sounds (and later pictures) via waves sent through the air; (5) *computers*, the programmed processing of information via wires (or optical fibers), wires that are microscopically small and integrated by the million on tiny grid platforms; and (6) *Internet*, the combination of inventions 1 through 5 in a globally linked network.

The first two of these inventions preceded the electrical grid. The last two came very late in the grid's history. Because the time frame in play in this chapter is the first third of the 20th century, we'll concentrate here on the fourth invention, radio, which was the next big thing to come along. Actually, radio didn't just come along. It wasn't a single invention or insight. It was built on clever ideas, scientific discoveries, and investments spread across the better part of a century.

To start with, the time from Faraday's flicker of recognition of the true kinship between electricity and magnetism to the development of practical electric generators—having the ability to crank out lots of power—was several decades. From the time of Maxwell's insight that electric and magnetic forces together conspired to constitute ordinary light (full technical name: electromagnetic radiation) and, furthermore, that light comes in many varieties (such as ultraviolet and infrared) that the human sensory system cannot directly apprehend to Heinrich Hertz's creation of one form of that nonvisual radiation (what we now call radio waves) was 20 years. From Hertz to Guglielmo Marconi's throwing radio waves across the Atlantic was a further dozen years. And it would be another couple of decades before regular radio broadcasts came into the home. Here's how it happened.

Radio waves are, first of all, an example of electricity and magnetism mixed together. The waves emanate from moving charges, but once the waves are created they're a separate thing, like sparks flying from a grindstone. And once set loose, the waves leave the scene at the speed of light; indeed, they *are* light. They can, under the right atmospheric conditions, even reflect off the undersides of clouds and thereby travel a considerable way around the curve of the earth's surface.

To be useful for mass communication, however, the tiny electromagnetic disturbance needs to be amplified and manipulated using special circuitry. To make the transition from a *radiant thing*, a wave phenomenon moving through the air, to a *circuit thing*, a small-scale burst of electricity moving along metal wires, and then finally into an *acoustic thing*, a train of sound waves crossing the room, a whole new arm of the electrical industry had to come into being: electronics.

It was in the early radio years that the particulate nature of electricity, in the form of electrons, was discovered. In 1897, J. J. Thomson in England built himself a glass tube containing two electrodes at opposite ends, one hooked up to the positive terminal of a battery and the other one to the negative terminal. Even with no wire stretching between the electrodes, and with all the air in the glass pumped out, if the negative electrode was heated up, a current would flow across nothingness toward the positive electrode.

Years earlier Edison saw this odd behavior (now called the "Edison effect") but could make no sense of it. But Thomson did. From his simple laboratory setup several important facts were learned. For one thing, an electrical current consisted of tiny lightweight particles—electrons. The first true elementary bits of matter to be detected by scientists, electrons were lighter by far than atoms would prove to be. Moreover, the tube with the electrodes could be modified to function as an important circuit component, a "diode," that allowed current to flow in one direction but not the other. Diodes could be used to rectify current, from Westinghouse into Edison—that is, from AC into DC.

By adding a third electrode to the glass tube, the current leaping through vacuum from the negative (cathode) to the positive (anode) electrode could be controlled with great sensitivity. The whole thing, miniaturized and called an electron tube or a vacuum tube, became the central component in electronics, to be replaced years later by transistors and crammed by the million onto chips the size of a breath mint. Indeed, the miniaturization of transistors has proceeded so magnificently and the use of electronic products has become so pervasive that Gordon Moore of Intel Corporation has estimated that in one year more transistors are made than the number of raindrops falling on the state of California.[14]

Besides acting as a switch or a gate or an amplifier, the electron tube (eventually transistor) can be used as an oscillator: it helps convert electrical impulses from one frequency to another, from the

60-pulses-per-second time scheme of the electrical grid up to the hundred-million-pulses level of radio waves, or back down again to the thousand-pulses-per-second time scheme for acoustic waves. Pretty handy: one device that can adapt to the very different pace of the grid world, the broadcast world, and the audio world.

By the early 1930s nearly every home in Chicago had a radio unit. Samuel Insull was interested from the start. He could see that so great was people's eagerness to have radios that some of the last holdouts signed up for electric service in their homes just so they could receive the broadcasts that all their neighbors were talking about. It can be said, not so incidentally, that the possibilities for advertising products in conjunction with radio programs that penetrated thousands and millions of homes were not overlooked by certain enterprising individuals. And thus began the dubious epoch of broadcast ads.

As soon as the full sound of the human voice or a polka band, and not just the prosaic dot-dash of the telegraph, could be conveyed on the back of an invisible wave and sent hundreds of miles through the air, radio could fully break out into public use. Adventuresome local "stations" could set up and broadcast to a fraternity of nearby listeners who had built themselves receivers. Weather and crop forecasts, crime reports, Bible school, an up-and-coming local fiddler, a football game in progress: these were things that would make people gather 'round and listen. With time the programming got more ambitious; stations started broadcasting for longer periods over greater distances. The first stations with regular programming delivered the results of the 1920 presidential election: Harding beats Cox.

The Radio Corporation of America (RCA) built a network of these stations, becoming the National Broadcast Corporation, or NBC. In effect a radio grid was coming into existence, an information-laden, higher-frequency appendage of the power grid. The two grids were intertwined, and the growth of one helped the growth of the other.

How did electric machines change things? In the catalog of energy-transforming devices, incandescent-bulb lighting had come first, changing the way people used the night and day. The next to be institutionalized were electric irons, which also altered the rhythms, or at least mitigated the sweat, of housework. And what a salubrious effect this would have on the woman of the house! One not-so-subtle ad for electrical appliances got right to the point. It asked, "How Long Should a Wife Live?"[15]

The third big electrical best-seller, radio, changed the way families approached their evening entertainment. Previously, if you wanted music, you would sit down at the piano and play. Hereafter it would be easier to produce Beethoven by deploying a wax cylinder on a machine or by tuning in to the Philharmonic broadcast. The sale of sheet music declined. The keys were covered, and in some cases the parlor piano became a mere ornament.

THE PRICE OF ELECTRICITY

No more piano? Had it come to this? Lewis Mumford was worried. A writer at the *New Yorker* magazine for 30 years, he was one of the great culture critics of the 20th century, reviewing buildings, highways, machines, and whole cities the way a drama critic reviews plays. In fact, to him the march of technological development *was* a drama much like the Medieval morality plays.

With a firm grasp of history, Mumford was skillful at putting things in perspective. To depict this volcanic moment in the development of the grid, in the 1920s, just when electricity was about to spill into so many homes, perspective is what we need. Earlier I paired Michael Faraday with Henry David Thoreau and then Thomas Edison with Henry Adams in order to examine both the science-inventing and the social-philosophical sides of machine culture. Now I want to match the industry-minded and efficiency-minded Samuel Insull with the historically minded Lewis Mumford.

No more piano? Mumford was often bothered by something or other and devoted his writing career to telling you why. He was concerned with what he saw as an erosion of humanistic values, partly as a result of society's pursuit of an ever sweeter, mechanized efficiency:

> . . . A good technology, firmly related to human needs, cannot be one that has a maximum productivity as its supreme good: it must rather, as in an organic system, seek to provide the right quantity of the right quality at the right time and the right place for the right purpose. . . .

> The center of gravity is not the corporate organization, but the human personality, utilizing knowledge, not for the increase of power and riches, or even for the further increase of knowledge, but using it, like power and riches, for the enhancement of life.[16]

Not strive for maximum productivity? Not, Mumford argues, if it interferes with the important things in life, things like family togetherness, or being able to see the Milky Way or the ability to breathe fresh air or play the piano.

We're back to the Faraday–Thoreau dichotomy. The grid is a good thing. It lights our homes, preserves food, brings us cheap aluminum. It also brings us pollution, overhead wires, clockwork tedium, and a massive contribution toward greenhouse warming and climate change. Like Henry Adams, Lewis Mumford wants you to know the price of technology. He figures that if someone like him doesn't tell you the price, no one else will. Insull wants you to buy power; he's not going to tell you about the ancillary costs. Government regulators just want the marketplace to run smoothly. Bankers want to extend you loans; they'll leave it up to you to decide where the money goes. Cities want a wider tax base. Manufacturers want to sell you products whether you need them or not. Ad executives want you to pay attention to their pitch. Who among this cast of characters in the great morality play of public life will give you unvarnished advice? That's the job of people like Mumford. It's also the reason behind my own modest expressions of cheerful ambivalence about things electrical.

The beginning of the machine age, Mumford will tell you, began not with the grid or with steam power but with the invention of the mechanical clock in the 13th century. Edison, with his bulbs, divided light into small parcels. Westinghouse, with his transformers, could divide volts, and Tesla divided torque—he made motors flexible enough to run small sewing machines. What the monks of the 13th century did with their clocks was to divide the hour into minutes.[17] Time was no longer a fluid thing but something palpable and divisible. Starting with the monks themselves, with their religious ceremonies and chores dictated by a strict hourly schedule, life became much more regimented around time. Thereafter, people would increasingly eat by the clock, not when they were hungry, and sleep by the clock, not when they were tired.

No more piano? Mumford could see that the electrical grid's impact was not all good. Still, he hoped (writing around 1930) that the worst aspects of the Industrial Revolution (what he called the paleotechnic phase of history) would be mitigated by the coming of the electrical (neotechnic) era. He hoped the grime and noise and tedious work drill of an age characterized by coal, steam, and steel, an age so well depicted by Charles Dickens, could evolve into an age characterized by a quieter and cleaner energy source, supplied from afar by electricity. This evolution, according to his city-planning perspective, would encourage a dispersal of the high-density living and working concentrations. Smaller and more livable cities would result.[18] We'll see later whether Mumford's hope was fulfilled.

MAXIMAL GRIDIFICATION

Those who suppose that the intensity of technological change during their lives will never be equaled are probably wrong. Compare the 1990s with the 1890s. Consider, for example, the 1990s' electro-tech innovations having a large worldwide impact: ever more compact personal computers, wireless communications, e-mail, the Internet, and optically read compact discs for music and video storage. An impressive list. But look at the multiple inaugurations of 10 decades before: radio signaling, diesel engines, motion picture cameras, x-rays, AC induction motors, radioactivity, and automatic telephone switchboards. One could probably locate many decades when the comparative change in technological life was just as noticeable.

The 1920s were such a time. Now that the price of electricity (the cost of making the power, anyway) was low enough, the lid was about to come off. Cities in most places were now rapidly installing more wires, more power lines, and more radio antennas than ever before. The most aggressive bustle was in Chicago. The 1920s were a roaring decade. The Great War was over and the Windy City was one of the largest urban centers in the world, partly because of the jobs that come with booming productivity. And a big part of this great leap forward was powered by electricity. Here are the numbers. The electrification of Chicago's factories between 1900 and 1930 went from 4 to 78 percent.[19] In that remarkable 30-year period the rates charged for power fell by half even as the price of the fuel used to make electricity tripled.[20]

Compare Chicago with other cities. In 1912, electric sales per person in London were 49 kilowatt-hours per year. In Berlin it was 83 and in Chicago a whopping 291 kilowatt-hours. In London a kilowatt-hour cost 4.8 pennies, in Berlin 3.9, and in Chicago 2.2. What you paid depended at least in part on how big your local generator was. Bigger dynamos generally made electricity more cheaply. In London the average generator size was 5 megawatts. In Berlin it was 23 and in Chicago 37 megawatts.[21]

On an atlas of worldwide electrical cultivation, the brightest zone, the very most intense buildup, would center around Chicago in the decade from 1918 to 1929, when a majority of homes in the city were wired up.[22] Chicago was the most electrified city in the world, with an average per-capita annual consumption by 1925 of nearly 1,000 kilowatt-hours.[23]

No more piano. What was the need? People listened to the radio in the evening. What other changes had been ushered in? In 1915, 1 in 60 families had cars in the United States; by 1925 the number was 1 in 8.[24] And of course that fraction would grow further. One victim of this latter development? Trolley ridership. The car culture was being launched.

Thomas Edison went so far as to suggest that electrical devices would accelerate human evolution. In times to come, he said, a woman, executing her home duties, would be more like "a domestic engineer than a domestic laborer, with the greatest handmaiden, electricity, at her service."[25] A wife's brain would come to equal that of her husband's.

To say that it didn't work out that way in practice would be an understatement. For one thing, progress can go in reverse. Doing the family laundry, for example, had been done by hand, an ordeal that could take many hours. Later, for many middle-class households, the wash was then given out to a professional laundry. Later still, with the arrival of electric home washing machines, the dirty clothes again stayed within the home. Net result of electrification? More labor than ever for the "domestic engineer" in the family.[26]

For another thing, evolution usually doesn't work that quickly.

There is always a great lag. Squirrels, for example, know how to avoid predators in the field, but they haven't wised up to automobiles. Humans themselves, for all their ingenuity and machines, still basically possess the brain structure they had when the species was limited to hunting and gathering on the East-African savannah, as current theories of anthropology would have it.

Had electrification come too quickly? Are we squirrels about to be run over by our own machines? Our civilization is older, but is it wiser?

ASSYRIAN EMPIRE

The grid had arrived, at least in many major cities. It was in people's homes. Samuel Insull had a lot to do with that. To achieve his ends he had sought and achieved a monopoly over grid activity. If you lived in Chicago and wanted to be on the grid, you had to come to him. His rates, nevertheless, were lower than almost anywhere else, including New York, Philadelphia, Boston, and Baltimore.[27] For these many achievements he can be considered a hero of the grid. This is the Good Insull.

We will also now have to review what history has come to see as the Bad Insull and to explore why in general his reputation, if he is recognized at all, is under a perpetual cloud. Insull's lone biographer, Forrest McDonald, sees Insull not as a greedy or evil man but as one whose main fault was the pridefulness to believe that he always knew more than other people.[28] Past experience had taught Insull that he could solve any problem; he was better at organizing large companies or cajoling politicians into giving him what he wanted; he was shrewder at discovering cheaper and more reliable ways of getting the raw materials he needed for feeding his grid; he was abler than other merchants in procuring new customers, customers he dearly needed for smoothing out his load curve.

Arguably he was better at running other people's businesses than they were. He helped fix and improve the Chicago transit system. His greatest performance in a supporting role was his action in rescuing Chicago's insolvent gas company. Insull's credit, and the credit of his home company, Commonwealth Edison, had the strength of granite. Investing in Insull was a sound investment. His companies always paid dividends. As a boss he was benevolent. The wages he paid were better than those at other utilities. His employees were encouraged to contribute time and money to charitable and civic causes. Often the senior managers were leaders in community activities. He offered his workers affordable life insurance, and the company owned a resort in Wisconsin for family vacations.[29]

What's so bad about all of this? Where is the *bad* Insull? Shouldn't crooks be made of meaner stuff? Compare, once more, the life of Samuel Insull and the protagonist of *Citizen Kane*. Kane bought up all his rivals, and so did Insull. In the movie the newspaper baron marries a singer and builds an opera house in Chicago. In real life the utility baron married an actress and built the *actual* opera house in Chicago. In the movie we see Kane performing noble deeds and pulling the levers of power. He wants the public to love him and he enjoys influencing events. The same with Insull. He didn't need to be mayor to be Chicago's premier citizen. The opera house, looking somewhat like a high-backed chair, became known as "Insull's Throne." "I am not a musician," he said, "nor am I in any sense an authority on grand opera, except as to what it costs."[30] Is it any wonder that when Orson Welles was transforming himself into the elderly Kane, the photograph he kept on his makeup table showed the imposing dome head of . . . Samuel Insull?[31]

Like the Assyrian empire, Insull's fief had started small. Then with dynamos as his sword and patents as his shield, he conquered the rest of Chicago. Next was the North Shore Electric Company, a confederacy of grids lying along the rim of blue Lake Michigan as far as the Wisconsin border, as well as a few petty dukedoms to the south. Insull secured sovereignty over the northern regions by tethering them to a 138-kilovolt transmission line, as practical and potent a symbol of imperial sway as the Roman aqueduct or the Chinese Great Wall had been 20 centuries before.

He had bought suburban and transurban grids wholesale. Insull's domain now covered thousands of square miles. Confident in his vision, Insull's transmission lines even went where previous grid builders had feared to tread. If he had heard there were potential customers on the moon, he probably would have found a way to get there and lay wires in the lunar dust.

Insull encouraged his employees to buy stock in their own company, as he had done himself. They should buy stock and their friends should buy stock too. You, the employee, could earn a small commission even as you helped your friends. All would benefit from the health and advance of the company. What's bad about that?

With the advent of Middle West Utilities, Insull's reach grew to the continental level. This corporation was a holding company. It advanced money or equipment or patent licenses to struggling new utility companies, which in return gave stock. The early Edison company operated along these lines—the new grid affiliate in Cincinnati, say, could hardly afford to buy equipment *and* operate as a grid without surrendering some ownership to the larger company holding the patents and making the machinery. The holding company, better known and better trusted than the subsidiary operating company, could then sell *its* stock, using as collateral the stock held in the operating company. Insull came to control utilities all over the place—California, Ohio, Kentucky, Louisiana, New England.[32] The grid was now virtually everywhere, and Samuel Insull had done much to make it what it was.

To mark the gigantic transformation in society that electricity had made, a special occasion was being observed. October 1929 was the 50th anniversary of Edison's development of a practical lightbulb. Henry Ford had built a museum to honor Edison, and the centerpiece was a painstaking reconstruction of the wizard's Menlo Park lab complex, the scene of Edison's greatest research as well as his enchanting encounter with Sarah Bernhardt and his dinner with the aldermen of New York. The resurrected lab was built at Greenfield Village, not far from Ford's mammoth auto factories in Detroit.

Edison, now 82 years old, was lauded by the president of the United States, Herbert Hoover. Edison and one of his original assistants staged the lighting of the 1879 bulb.[33] The reenactment of the historic moment of 50 years before was itself historic because it was performed before a transoceanic listenership of millions. The relighting of the bulb, narrated anew as if it were a championship sporting event, and the remarks made by Edison and by President Hoover, were beamed across the United States and to other lands.

Beamed in the opposite direction came greetings from Germany, from President Von Hindenburg and also separately from the famous Albert Einstein. From the Antarctic came felicitations from explorer Robert Byrd. The domestic part of this "greatest hookup that radio has yet attempted" was an ensemble of more than 130 stations coast to coast, beating the previous record of 111 stations.[34]

More memorable was what happened three days later: the greatest stock market crash in American history. The economic downturn of the early 1890s had led to Edison's removal from control of his company. The great market tumult of 1907 did the same to Westinghouse. But these events were small in comparison to the crash of 1929. This notable episode would soon touch the entire nation and other countries too.

Consider, as part of this vast turn of events, the plight of utilities. When suddenly the price of many stocks fell simultaneously, companies that regularly borrowed huge sums of money on the strength of stock holdings were obviously going to be nervous. Insull's empire encompassed utility operations in 32 states. With his index finger he controlled the flow of about one-eighth of the nation's power. One million people, from laborers to millionaires, many of them enlisted through the grassroots level, were investors in Insull's companies.[35] They all waited to see what would happen. A week after the crash Insull's face was on the cover of *Time* magazine. Against terrific pressures he kept his operations solvent, but many people were worried. Already the issue of public versus private ownership of utilities had become a major point of debate in American politics. Insull was seen as the leader of what had come to be called the Power Trust.

The utilities, and Insull in particular, were seen as unfairly influencing political affairs. Insull had a habit of making generous campaign contributions, sometimes to both parties. Entangling himself in a 1926 senatorial race, he had outdone (and maybe undid) himself by giving a reported $125,000 to the Republican candidate. Although this contribution was not illegal, the size of the gift and other factors at the time provided fuel to those who saw this as a grab for political power by the purveyors of electrical power. In our time the argument continues over what constitutes a principled political endorsement or an outright bribe. Franklin Roosevelt, in his 1932 presidential campaign, singled out Samuel Insull by name as being interested in acquisition and not service.[36]

By June 1932 too many holes had sprung in the dikes, too many loans were being called in, and Insull was forced to surrender his management positions by the New York bankers, the people whose clutches he had always tried to elude. He even had to give up his beloved Commonwealth Edison. Insull sailed for Europe and moved into a Paris apartment. Then worse: He was indicted on charges of fraud, embezzlement, and other crimes related to his Byzantine holding company.

Leaving for the next chapter the question of his guilt or innocence, we now concentrate on Insull's escape. The conqueror of the known electrical world had become a fugitive. He fled to Greece, where there was no extradition treaty with the United States. There was even a possibility that he would assume Greek citizenship and be put in control of the national grid there.[37] Insull, in his memoirs, tells of an even more poignant what-if scenario. Years before, Prime Minister Stanley Baldwin had secretly invited him to return to Britain and take charge of the commission that was to reform the British national grid. After giving the matter serious consideration, Insull regretfully declined. "It seems my life would have been pleasanter and both myself and others would have been saved lots of trouble if I had accepted Mr. Baldwin's offer."[38]

He hadn't accepted the offer but stayed in the American arena, and why not? He had been present at the birth of the grid, standing at Edison's side on Pearl Street. Insull had gone on to fashion his own powerful grid, the Chicago grid, a million times bigger than Pearl Street. But he never forgot his origins. Insull's employees referred to him respectfully as "The Chief," but for Insull himself there was only one chief. He idolized the Old Man of Menlo Park. In the bound edition of Insull's most famous lectures, the frontispiece illustration is a photograph of Thomas Alva Edison.[39]

There is also the matter of that boast, the promise made at the banquet sending him off from New York to Chicago 40 years before, a vow to outdo General Electric. It was an assertion made half in jest and half in earnest, but it does seem to have captured the essence of the man and what he subsequently did in Chicago. Little Sammy Insull, Edison's secretary, had gone on to the big time. Accepting custodial duties over Britain's grid would have taken him away from a culture where governors and senators were his friends. Whole cities had electricity because of him. He had done something worthy of Thomas Edison.

Many other great men in history had suffered great reversals of fortune. Napoleon had conquered all of Europe as far as Moscow and then was himself conquered, his empire shrinking from the size of a continent to the size of his jail on St. Helena Island in the south Atlantic. Agamemnon, breaker of cities, supreme leader of the Greek army at Troy, returned home, where his rulership constricted with lethal rapidity. Stepping from a ritual cleansing bath he was axed to death by his wife.

For Insull it wasn't quite as bad. Leaving Greece and heading for Turkey, he was finally apprehended through the vigorous efforts of the U.S. government. Shanghaied from one ship and placed in another and suffering weeks of blazing front-page newspaper coverage, he was ignominiously transported back to New York, driven to Chicago, and remanded to a jail cell, where he spent the night. In the breadth of his career, the magnitude of his influence, and the height of his fall, Samuel Insull's saga could almost be compared to Greek tragedy, and somebody someday should write a play about it.

4

IMPERIAL GRID

Ever curious and on the lookout for a dramatic story, whether for his novels or his newspaper articles, H.G. Wells was shown into Vladimir Lenin's presence. Wells, author of such books as *The War of the Worlds* and *Time Machine*, was himself a notable personage and had found it tedious to have to negotiate numerous sentry points before penetrating to this inner office. But the quality of the interview with the Bolshevik leader more than compensated.

Wells had expected to meet Lenin the doctrinaire Marxist, Lenin who was known for his frequent cynical laugh, or the Lenin who liked to lecture people. Instead Lenin was none of these. He did, however, live up to his reputation for intensity.

Soviet historians liked telling the story of Wells's 1921 visit to Moscow, his inspection of the ruined economy, his lively interview with Lenin, and his gloomy prognosis for the coming years. Wells was an astute observer of many things, and so his predictions concerning the proposed electricity-centered industrial revival in the Soviet Union were particularly troublesome. Basically Wells saw Lenin as an impractical dreamer:

> Can one imagine a more courageous project in a vast flat land of forests and illiterate peasants, with no water power, with no technical skill available,

> and with trade and industry at the last gasp? But their application to Russia is an altogether greater strain upon the constructive imagination. I cannot see anything of the sort happening in this dark crystal of Russia, but this little man at the Kremlin can; he sees the decaying railways replaced by a new electric transport, sees new roadways spreading throughout the land, sees a happier communist industrialism arising again. While I talked to him he almost persuaded me to share his vision.[1]

Lenin suggested that the skeptical Wells return 10 years later to see if the vision had been carried out or not. One can almost get the feeling that the subsequent history of the Soviet grid was geared toward wiping the condescending smile off Wells's face. Dark crystal? They would show *him*.

FUEL FOR RED OCTOBER

Was electricity a service to be provided to citizens, like drinking water and roads? Or was it a product, like shoes, to be sold at a price determined by competition? These questions, asked over and over wherever an electrical grid was set up, were especially interesting in the Soviet Union. There the grid was neither product nor service but rather a weapon, a tool, to be used by the state authorities to strengthen the nation, defeat its enemies, enlarge the economic base, and further the revolution.

In Tsarist Russia the grid had not gotten much beyond the early for-the-rich-only phase. Electric lighting had arrived early—some of Edison's bulbs had burned at Alexander II's coronation in 1881. After that, though, growth had been sluggish. Lying at the eastern geographic extreme of what counts as Europe, Russia was often off the map when it came to trade, finance, and scientific and technological innovations. Government involvement was either repressive or, just as bad from the standpoint of encouraging development, neglectful. Huge import duties, illicit payoffs, delays, indifference, mistrust between government ministries—Russia was not the kind of investment environment to attract the big money needed for building an energy infrastructure.

Here's where things stood on the eve of the Bolshevik revolution. Electrification was 20 years, maybe more, behind the other parts of Europe. In and around the two major Imperial cities of Moscow and St.

Petersburg, the grid depended on elderly generators of Belgian origin. As for fuel, coal was brought all the way from Britain or hauled from districts in Russia so far away that they might as well have been abroad. Electricity cost four times more in Moscow than it did in Amsterdam, and electrifying a tramline in Omsk meant that no new customers could be added to the grid for years.[2] Then things got worse.

The lore of nations emphasizes heroic deeds and oversized characters. Britain has its King Arthur and Henry the Fifth. The United States has George Washington, the father of the nation, and Abe Lincoln, the Liberator, while the Ottoman Turks have their Suleyman the Magnificent. The Soviet Union has Vladimir Ilyich Lenin, who is always portrayed as an earnest, idealistic but pragmatic, visionary in his designs and frantically eager to begin the task at hand. One of the biggest tasks in revolutionary Russia was the restoration and expansion of the electrical network following the devastating years of back-to-back wars, including World War I, the Bolshevik takeover, and the civil war between the Reds and Whites.

War and revolution, retreat and retribution, wrecked just about everything. The grid would have to be built from the foundation back up. And this time the way things were done would be very different. According to the glorious legends (and here, I confess, I find it difficult to pass up the temptation to write in the fervent tone of the times), the Commissars wanted to show the world what a communist grid could be like. No pooled ownership, no consortia, no state-regulated private utility, no bourgeois equivocating of any kind—nothing but full government operation.

Lenin, eager to reach boldly for anything that would advance the revolution, was tantalized by the transformative attributes of electricity. Consequently, he sent for the top electrical engineer around, Gleb Krzhizhanovsky. The animated discussion between the two of them, conducted by candlelight since electricity was scarce, was about fuel. There was no question of getting coal from capitalist Britain anymore, and fetching forth much coal from Russia's own Donets Basin far in the south was problematic considering the crippled state of the railroads.

Local fuels were the answer. "We must declare a proletarian crusade for peat," Comrade Krzhizhanovsky declared in the robust style of the time.[3] Bog world: Russia possessed an enormous carpet of potential energy in the form of its 240 million acres of peat. Peat, if you compress it in the earth for eons, is what becomes coal. But even taken as is, it is a usable fuel. Turn it over, stack it up, dry it out, send it through a furnace to make steam. This was Socialist ideology in action. Decisive, direct, strategic!

The next day Krzhizhanovsky, having hardly slept after spending a night conversing with the revered leader of the revolution, was startled by the arrival of a motorcycle messenger from Lenin. "Your report on peat was most interesting," the missive asserted.[4] Lenin wanted a more forceful text, something longer, inspirational, something to go into the newspaper *Pravda*, where it would arouse a larger audience. Krzhizhanovsky went to work and produced a letter, "Peat and the Fuel Crisis," which was duly published along with a letter from Lenin called "The Fight Against the Fuel Crisis." There ensued numerous other letters to Pravda debating the merits of local fuels.[5]

Soon another motorcycle message from Lenin ordered Krzhizhanovsky to bundle his various studies of the grid situation (hurry, please) into a pamphlet, which Lenin needed as ammunition for a speech he would make at the upcoming All Russian Central Executive Committee. With the greatest of urgency the required task was again accomplished. Just in time, printed on a crash basis, hundreds of copies of Krzhizhanovsky's report, accompanied by hand-printed full-color maps, were distributed to delegates.

Lenin quoted from the report to urge the necessity for electrification: "The age of steam is the age of the bourgeois; the age of electricity is the age of socialism." What came of Lenin's emphatic urgings was the creation of the State Commission for the Electrification of Russia. This organization, Lenin hoped, would be the vanguard of a movement allowing Russia to out-develop the bourgeois powers. The electoral world was to be another theater of the people's struggle.

Two hundred engineers shut themselves into a building and went to work. The kind of power network that in the West had evolved piecemeal over several decades, mostly through the actions of entrepreneurs like Edison and Insull, would in Russia be dictated by a government edict assembled over 10 months. Lenin ordered extra firewood to keep the commissioners warm during their work. He saw to it that they received Red Army food rations, better than ordinary members of the government received. Lenin phoned occasionally to see how things were going, to cheer them along and exhort them in their business. With V. I. Lenin looking over your shoulder you didn't stop for want of a salary or for family matters. What were petty inconveniences when the nation was waiting for their plan? Lenin was excited and therefore so were you.

Still another motorcycle missive arrived from the Kremlin. Lenin demanded more speed and spoke joyously of "melting church bells for copper wire" and "placing a lightbulb in every village."[6] The report had to be ready for the Party Congress. Already breakneck, the work pace accelerated. Once-in-a-lifetime exploits were at hand. Now was the decisive moment. If the events of the crash program were to be made into a film, the director would have to be Sergei Eisenstein. As a sound track he could reuse Prokofiev's martial score from *Alexander Nevsky*.

Done at last! Another blitz copying job was required. Five separate print shops around town were commandeered to produce the needed copies of the 600-page commission document, intended to be the nation's electrical blueprint for the next 10 years. Future ages will marvel at the audacity. (Further disclaimer: My mock-heroic expositional treatment of Bolshevik electrification should be juxtaposed with the grim facts of life in Russia in those years—many died before firing squads and many more of starvation and exhaustion in work gangs.)

The hypothetical grid film would now cut to a snowy city street in front of the gorgeous Bolshoi Theater. Delegates arriving at the Eighth All-Russian Congress of Soviets are questioned before entering by sentries of the Red Army. December 21, 1920, is Lenin's big day, and the absence of heat in the auditorium is a trifle. The esteemed participants will keep their greatcoats on. Parts of Moscow are blacked out in order to provide enough electricity to light the auditorium. Here he is. Lenin, coatless, is soon welcomed to the podium, where he will speak without notes, casting off that charismatic glow that sets him apart.[7]

His speech, and the purpose of the Congress, is much concerned with nation building, but it takes its place also in the roster of famous lectures on electricity. Faraday considered it his duty to inform the public, which turned out in great numbers to hear of his newfound mastery of electric currents. Edison, in a boastful little sermonette in 1879, had prophesied the universality of energy spreading across wires into people's homes. Tesla, in his 1888 lecture before a roomful of attentive engineers, spoke of a revolutionary motor that would henceforth do civilization's heavy lifting. Samuel Insull, addressing in 1898 his brother utility operators, not one of whom was anything short of a capitalist, made his mark by greatly enlarging the number of people who could use electricity. He did this by paying attention to load curves and rate making.

Now it was Lenin's turn to address a rapt audience—all professed anti-capitalists—and alter grid history. Lenin, arch conspirator, who had been in and out of jail, who had personally ordered the executions of many men and women, was now proudly standing before a roomful of his fellow revolutionaries, many of whom had fired a gun into a man's head or liberated a factory while holding a red flag. Lenin had been the chief architect of the main program of the Bolshevik Party. This First Program of the party had been to overthrow the existing order and set up a socialist government. Now, to the delegates of the new regime, he was promulgating a Second Program for the party, namely the vast enlargement of the Russian economy through electrification.

Lenin's most famous slogan was later emblazoned on the wall of every power station. The force of the remark and its formulaic, almost incantory, nature obliges the use of capital letters:

> *COMMUNISM IS SOVIET POWER PLUS THE ELECTRIFICATION OF THE WHOLE COUNTRY.*

Regardless of whether we believe eyewitness accounts that at this point in the speech listeners actually leaned closer to hear every word, or that during the technical presentation of the plan the revelation of each new power station on the display map was met with delighted murmurs and gasps, or that the voting members of the Soviet would always keep in mind that the man at the podium might devise a quiet liquidation for noncooperators, we should recognize the audacity of the proposed program.

The Soviet approach to gridwork far outdid Samuel Insull's in smoothing the load curve. In many places dynamos were put to efficient use around the clock. Scheduling did not separate Sunday from the rest of the week. Planners could dictate production and consumption. They could decree, say, an increase in the number of refrigerators by a certain factor for the coming year, and too bad if you weren't on the list. The aim was not merely to electrify the nation, not merely to surpass the Europeans, but to catch and surpass America. And they nearly did it. At the very least, they proved H. G. Wells wrong.

It wasn't easy. Starting from practically nothing the mighty grid came into being. Rail lines had to be built to the sites of some of the new power plants. In Petrograd the big power plant changed its name and its fuel type. The new name, "Red October," came from the overthrow in October 1917 of the interim government that had itself overthrown the tsar earlier in the year. As for fuel, in Tsarist times coal was fetched all the way from Cardiff, in Wales. During the Russian Civil War they had burned wood. Now, sensibly enough, the fuel would be peat from Mother Russia. The engineers hurled themselves into locating faster methods of digging up this fossil fuel, essentially solidified swamp. It lies right at the surface; retrieving peat is like rolling up a carpet. Up went carpet, out came electricity. Red October returned the favor by sending a portion of its output to energize the very harvesting machines that were bringing in the new peat.

Red October was still on duty when the German army laid siege to the city in World War II. By the time the tanks rolled up in 1941, the city was called Leningrad, in honor of the founder of the Soviet state. Lenin's passing in 1924 marked a turning point in Russian history, featuring the rise to prominence of Joseph Stalin and the rapid decline of Leon Trotsky. Trotsky, creator of the Red Army, found himself on the wrong side of many ideological issues and, more lethally, on the wrong side of Stalin's ambition. Following Lenin's death, Trotsky's comeuppance was meteoric. He quickly lost his post of war commissar and was given various lesser jobs, including the directorship of electrotechnical activities.

Trotsky was in effect ruler of the Soviet grid and, although he did not know much about electrical things at first, he turned with characteristic relish to engineering matters. "I was taking a rest from politics and concentrating on questions of natural science and technology." He read books, consulted experts, drew up reports, inspected power stations, and formulated recommendations.[8]

Unfortunately from the Kremlin point of view, and more particularly from the Stalin point of view, Trotsky was too important a personage and too great a threat to be allowed even the relatively peaceful and politically innocuous work of administering electricity. Once, after returning from a visit to a prospective hydroelectric project on the Dnieper River, Trotsky dutifully submitted an enthusiastic summary of the solid results attained, to which Stalin reportedly replied that "the project would be about as much use to Russia as a gramophone to a peasant without a cow."[9]

Having worked in the corridors of Soviet operations for so long, Trotsky had come to know his adversaries well and soon realized that he could never "take a rest from politics." Writing years later in his autobiography, he summed up his intolerable situation:

> The electro-technical board and the scientific institutions began now to worry them as much as the War Department and the Red Army previously had. The Stalin apparatus followed at my heels. Every practical step that I took gave rise to a complicated intrigue behind the scenes; every theoretical conclusion fed the ignorant myth of "Trotskyism." My practical work was performed under impossible conditions. It is no exaggeration to say that much of the creative activity of Stalin and of his assistant Molotov was devoted to organizing direct sabotage around me. It became practically impossible for the institutions under my direction to obtain the necessary wherewithal. People working there began to fear for their futures, or at least for their careers.[10]

Trotsky had earned his paranoia. In short order he resigned his electricity and engineering jobs, was relieved of all further government positions, was expelled from the party, exiled to a distant border town, and finally deported abroad. He lived in Turkey, then France, then Norway, and lastly Mexico. The days of this onetime giant of Soviet life, master of the army and of the grid, ended in 1940 when an assassin buried an axe in the back of Trotsky's head.

PROGRESSIVE GRID

This could not happen in the United States of America. Not the liquidation of a prominent leader and not the creation of a socialist grid. At least most people would have thought so. So great is the American anathema against anything nationalized that responsible officials, public or private, business or government, could probably not call for a national energy network, at least not one under the technical ownership of the federal government. True, the government had helped build some giant dams out west, such as the Hoover Dam on the Colorado River and the Grand Coulee Dam on the Columbia, but the electricity flooding from these projects was sold to outside utilities, which in turn sold it to customers. This, American citizens would be reassured to know, was not socialism. Nevertheless, the U.S. government *was* about to get into the electricity business in a big way, owing to the confluence of certain economic and political factors.

In the 1920s, just when Samuel Insull and the Power Trust were riding high, some members of Congress thought that the utilities were too powerful and exercised too much influence over public life. Without actually saying that capitalism was bad or intimating that it was immoral to run a corporation solely on the basis of turning as large a profit as possible, certain reformers wanted to clamp down on what they saw as destructive greed.

The issues of whether electricity is a service or a product—it's a bit of both, most would concede—and whether it should be made by public or private utilities would come to a head over the matter of what to do with the rich but underdeveloped hydroelectric resources of the Tennessee River. The Depression had deepened, Insull's corporate framework was dismembered in a spectacular manner, and utilities in general were beset with debt. Most importantly, a man itching for reform was elected to the White House. The new president had a long record of rancor against the producers of electricity. As a state senator he had tried to get New York to supervise development of the Saint Lawrence River for producing power. As governor he had sought to curb the rates and influence of the utilities. And as a presidential candidate he had railed against the holding companies and had singled out Insull for criticism.[11]

With the exception of George Washington's first term, or the embattled tenure of Abraham Lincoln, no presidency had gotten off to a faster start than Franklin Delano Roosevelt's. The so-called Hundred Days saw the passage of many major pieces of legislation, including in May 1933 the creation of the Tennessee Valley Authority, or TVA. The federal government, after so much reluctance, was finally in the power business.

How unusual was the TVA? It was established as "a corporation clothed with the power of government but possessed of the flexibility and initiative of a private enterprise."[12] Beyond this, the exact programs of the TVA, its methods, its goals, and its building projects were largely left to a triumvirate of directors. Two of the three FDR chose had been college presidents; one had an agricultural background, the other expertise in flood control. The third director, a man of only 33 when he took the job, will receive all the attention here because he was the person most associated with the TVA as it became a force in the lives of Tennessee Valley people. He would also become a prominent figure in establishing the ways in which the federal government would relate to private utilities to this day.

FDR's 1933 crash program to redirect the efforts of government to inspire the citizenry and invigorate the economy resulted in many programs with notable acronyms. The most visible were the Works Progress Administration (WPA), the Civilian Conservation Corps (CCC), the National Recovery Administration (NRA), and the Securities and Exchange Commission (SEC). Of these, the creation of the SEC and the passage of the Public Utilities Holding Company Act (aimed partly at companies like Insull's) had a considerable immediate effect on the way large utility trusts were configured. But of all the New Deal programs, perhaps the one that had the greatest impact on the way electricity was made and delivered would be the TVA, especially in the hands of the young director who, I shall argue, should take a place in the electrical hall of fame, not for his inventions but, like Samuel Insull, for grid management innovations on a large scale. The age of the patriarchs might have closed out, but great grid developments still lay ahead.

A short history of the TVA, then, will constitute the next act in this drama of volts. The TVA was not just one of many New Deal programs but the epitome of what FDR thought government could accomplish and the essence of what he was trying to do in the American heartland. It was, he said, "the apple of my eye."

BIBLICAL PROPORTIONS

David Lilienthal was not Samuel Insull. Insull, immigrating from London as a young man, had not gone to any proper college but instead had learned the electrical trade in the most practical way imaginable, by working for Thomas Edison. Lilienthal, by contrast, was a product of farm belt Indiana and Harvard Law School. Insull came to Chicago in 1892 at the age of 32. He made his name building a vast grid empire. Lilienthal came to Chicago 30 years later at the age of 24. He made *his* name as a lawyer, first in labor disputes, then in utility matters (including one case against Insull), and later in politics at the state, national, and international levels.

They're paired here, Lilienthal and Insull, because despite their great differences—the one a conservative, the other a liberal; one a businessman, the other a government official—they had much in common. Insull is underappreciated in the social and technological history of America and Lilienthal even more so. Insull achieved his importance by vastly enlarging the layout of the grid in Chicago, making it the most electrically intense city in the world, and also by widening the universe of human activities served by electrical energy. Lilienthal also achieved fame by going to a place where previously only small amounts of electricity flowed and then building a mighty grid, a grid consisting partly of wires and partly of government statutes. He too helped widen the scope of electricity both by making it tangible in the lives of poor people and by reinventing the role of government in energy transactions.

While still in his 20s, Lilienthal was regarded as one of the top lawyers in the country on utility matters and was often consulted by prominent people in Washington. So it happened that in 1930 when the new governor of Wisconsin, who had come into office on the same reformist tide that two years later would carry FDR into the presidency, needed to appoint a state commissioner who could do battle with local utility companies, he knew where to look. With a mandate to lower rates and curb the influence of electric and telephone companies, Lilienthal took to his new job of public service commissioner with gusto.

Newly arrived in Madison, one day he was intrigued to hear that a utility matter was to be argued that day before the state supreme court. Going along to watch, he chatted with the assistant attorney general, who quickly saw that Lilienthal was much more knowledgeable than he was on the subject. Lilienthal read the brief, took over the case on the spot, made the arguments, and won. "It was a thrilling and satisfying experience," he wrote in his journal. He wanted more of the same.[13]

In Wisconsin, as in many places in the United States, the local electrical utility was typically a private business operating as a monopoly. In that case, you ask, why can't the utility charge whatever it wants? Because the utility would have been regulated by a city or state agency. This is the way most utilities had done business since the time of Edison and Insull. The method for arriving at rates was to grant the company a certain profit on top of its expenses. Exactly how much profit and what counted as an expense were two of the matters that regulators were paid to investigate. How much should electric power cost? This is often difficult to say. The cost of an apple is pretty much set by the free market. Stores charge as much as they think they can get, making due allowance for the store across the street that also sells apples and for the customers' option to not buy apples at all if the price is too high.

With electricity it's different. You can buy fewer or more efficient appliances, but mostly you have to go on using electricity. You wouldn't say to yourself, "Alright, today I'm not going to use any electricity," in the casual way you would say, "Today I'm not going to buy an apple." In addition, there is usually no competing utility across the street willing to send current into your home at a lower price, although this might change in coming years. Until then most consumers are stuck with the utility they've got. Hence the need for regulation.

Even with regulations, establishing the fair value of electricity isn't easy. Is it a product or a service? How much does the utility spend making power? In the 1920s some U.S. senators were bothered that the prices charged for electricity by Hydro Ontario on the Canadian side of the Niagara border were much lower than those charged on the

American side.[14] Franklin Roosevelt was also bothered, which is why he had wanted New York State to get involved in developing the energy resources of the Saint Lawrence River.

As president, Roosevelt could now finally conduct his experiment in power production and pricing. The Tennessee Valley Authority would try to integrate the best federal and business methods. Roosevelt's testing ground would be a watershed region of the Appalachian Mountains, his tool would be the TVA, and his point man in the enterprise would be David Lilienthal, an avid litigator known for his lawyerly skills and his crusading verve, just the qualities needed to sustain a project that would succeed or not depending on success in a number of arenas: the courts, congressional hearings, and hundreds of small-town meetings in the affected communities to discuss the merits of the plan.

For being FDR's favorite New Deal program, the TVA had been rather loosely defined. Some argued that it was a gigantic welfare project. Others thought of it as a massive, concerted government effort to jump-start the economy of the communities adjoining the Tennessee River. This is one of the great rivers of the American interior, a waterway wandering from headwaters in the mountains of eastern Tennessee, Virginia, and North Carolina, and then moving west and south into Alabama, before finally heading north up through Tennessee and Kentucky, where it meets the Ohio River at Paducah. This geographic region, an emerald realm the size of England, was singled out by President Roosevelt for special treatment. What that treatment would be had not been written explicitly into the legislation. TVA's mission, in generic terms, was to produce the greatest good for the greatest number of people.

The TVA *was* a welfare program, the most extensive and coordinated such project in U.S. history. It set out to restore forests, build dams, counteract the loss of soil, improve navigation, promote the manufacture and use of fertilizers, mitigate health hazards such as malaria, and even bolster local cultural activities such as the construction of libraries and the development of adult education classes. The TVA helped farmers and businessmen set up new agricultural ventures. All of these tasks had in the past been addressed singly by state or federal authorities but never before by a unified governmental approach. The TVA was to be a great social experiment, and if it succeeded, maybe it would be tried elsewhere. The president had often said that if things weren't working—and in Depression-era America, there was plenty of failure—new methods had to be tried.

The TVA proceeded to carry out these tasks. It changed the lives of millions of people in the Tennessee Valley, mostly for the better. But the TVA is best known for being a producer of electric power. The TVA built a lot of dams that filled up the riverine system of Appalachia with a succession of massive concrete walls, each with its own gigantic reservoir behind it. Each dam construction project had its poignant saga of work camps, home clearance, evacuation of households below flood level, and the creation of new ecosystems, replacing what might have been a sluggish stream with an extensive lakeshore. People had to move and cemeteries had to be relocated. One man, claiming that the flame in his family fireplace had been burning without stop down through the generations for a hundred years or more, agreed to move only if the fire, continuously lit, could be moved too, and so it was.[15] Whole towns were covered by the gathering waters in what could be seen by some Bible-minded skeptics as a miniature Noah's flood. In reality, Norris Lake, one of the new artificial bodies of water, extended for 47 miles: This was long but not exactly of biblical proportions. The collective impounded water covered a total area the size of Rhode Island.[16] Birds were obliged to nest elsewhere, fish were presented with new habitats, roads were rerouted. The Tennessee was no longer a mere river. It had become the "Great Lakes of the South," with associated shipping ports and recreational parks and resorts.

New homes went up. Old homes got new paint. A literacy campaign went forward. What got most of the scrutiny, however, was the electricity created as billions of gallons of water pushed past turbines on their way through those dams. Skirmishes erupted over how much fertilizer to use or how to keep down the mosquito population, but the big battles were over volts. Some had expected that, as in the case of Hoover Dam, the federal government would help out in the construction phase but not in the marketing and distribution phases. Many assumed that the energy produced at the dams would be sent out over wires and then sold to existing utilities (some private, some municipal), which would then deliver it to paying customers.

David Lilienthal disagreed. Harking back to the view that the waterway resource was not there merely to serve existing power companies, Lilienthal saw to it that those utilities did not necessarily take possession of the Tennessee's electricity. TVA just might market the energy itself. On this issue Lilienthal came into quick dispute with one of his fellow TVA directors, Arthur Morgan. Morgan was a high-minded man. *His* TVA would be an idealized version of public-private cooperation, government funds married to individual initiative, New Deal politics working seamlessly with traditional capitalism.

Lilienthal, also an idealist but more familiar with the pragmatic world inhabited by lawyers, was eager to joust with the companies. Lilienthal's TVA would be the means for determining once and for all the true cost of making electricity. With all the mechanisms of generation at its disposal, the TVA could furnish the yardstick for production costs that Lilienthal had never quite achieved in Wisconsin. No, if he had anything to do about it, TVA would not mutely surrender its electricity to outside utilities. The TVA would sell power to cities and factories directly. It might even build its own long-distance transmission lines.

The utilities screamed. TVA paid no taxes; they charged; TVA had access to guaranteed government loans; it had cooperation from the U.S. Army Corps of Engineers and other agencies. What TVA was doing was unconstitutional, they said. Improving navigation on the Tennessee was a proper federal prerogative. Reducing flooding through the use of dams was permissible. But making and selling a product, commercial electricity, was against what the founding fathers ever would have recognized as a proper government activity. Did the TVA represent the beginning of a socialist takeover? That's what the *Chicago Tribune* argued in an election-year editorial entitled "Vote for Republican Congressmen":

> Another Congress like this one which has authorized the NRA [National Recovery Act] to put small businessmen in prison, the AAA to destroy food, and the TVA to establish a little Red Russia in the Tennessee Valley, will do unimaginable damage to the American people.[16]

To get a sense of how radical an idea the TVA was, consider a modern day equivalent. Imagine the president announcing the creation of the Ohio Valley Authority, or OVA, an independent federal agency designed to rejuvenate aging rustbelt cities along the Ohio River, combining the functions of several existing government departments into a single unified effort. It would, for example, rescue disadvantaged inner-city schoolkids, teach parenting skills where needed, emphasize math and science in school, and steer students toward engineering careers. The real aim of all this social intervention, however, would be to pioneer a new form of energy transmission using hydrogen. OVA would construct a grid of separation plants, pipelines, service stations, and storage facilities in support of efficient hydrogen-gulping automobiles and superfast trains that traveled from Pittsburgh to Cincinnati to Louisville in mere minutes.

COMMAND OF RESOURCES

The utilities eagerly took TVA to court. Lilienthal was ready. He had plenty of utility litigation practice in Chicago and Wisconsin. The Tennessee Valley would be Wisconsin times 100. Lilienthal was not anti-business, not against private enterprise, he insisted, but he did believe he had received a mandate from the president, as he had earlier received a mandate from the governor of Wisconsin, to win lower electric rates for consumers and to thereby increase overall consumption, particularly among the disadvantaged, who had previously been left off the grid. And there were plenty of disadvantaged in the Tennessee Valley. Lilienthal wanted to be their champion. He was in effect a territorial governor operating at the energy frontier. TVA was his Wild West.

He explored ways of producing cheaper power and selling cheaper appliances. Indeed, by twisting some arms, Lilienthal got the major manufacturers such as General Electric and Westinghouse, to build appliances that sold for roughly half the previous price, thus encouraging a new era of electrical conveniences in the home.[17] Lower rates and greater consumption: Lilienthal was, in a strange way, a federal version of Samuel Insull. FDR was so pleased with these efforts that he ushered in the Rural Electrification Administration (REA) in order to promote electric use among other have-nots in the rest of the country.

The utility lawsuits had been unsuccessful. Decisions all the way up to the U.S. Supreme Court had validated the constitutionality of TVA and its role as a distributor of power. TVA's grid was now the largest government-controlled power network in the country. Ten years into the history of TVA it was clear that Lilienthal had won.

He wrote a popular book, *TVA: Democracy on the March*, to explain the purpose and achievements of his vast undertaking. The first thing he did in the book was to deny that TVA was meant as a threat to private business or that it was the first step in a socialist plot to take over America. "The TVA experiment has been carried out under the existing rules of the game of American life. It required no change in the constitution of the United States. Congress maintained full control. Property rights and social institutions have undergone no drastic amendment."[18]

There was, moreover, much to be proud of. TVA jobs, Lilienthal reminded his readers, were awarded on the basis of merit. The TVA system consisted of dozens of dams, constituting the largest engineering feat in U.S. history undertaken by a single organization, the excavation and construction amounting to the equivalent of 20 Empire State Buildings. The internal lake and river shoreline created in the process was thousands of miles long and rivaled in magnitude the Atlantic and Pacific coasts of the country.[19]

In his writings and speeches, Lilienthal emphasized the multipurpose aspects of the TVA's work, such as its promotion of fertilizer and better navigation. Once, when the New Orleans water supply was threatened by a backflow of water owing to the slackness of the Mississippi River, TVA had helped out by releasing more of its water.[20] The genius of the dam system allowed for this contingency. A few levers were pulled by engineers in the Appalachian Mountains, and 1,500 miles away New Orleans got the water it needed. Inevitably, however, both Lilienthal and his critics were drawn back to the electricity issue.

Waxing philosophical about the lamps, refrigerators, and electric farm implements newly possessed by valley inhabitants and citing impressive superlatives, such as the fact that the TVA would eventually become the largest power-producing region in the United States, Lilienthal liked to promote the making of electricity as being practically an *ethical* undertaking:

> Such figures as these [statistics relating to TVA electricity] are more than figures; they have deep human importance, for this must be remembered: the quantity of electrical energy in the hands of the people is a modern measure of the people's command over the resources and the best single measure of their productiveness, their opportunities for industrialization, and the potentialities for the future.[21]

The TVA had outlasted its critics. Its legal staff, called the best of any federal agency, had defeated the utilities in court, leaving TVA the sole power producer in the valley. Some argued that the TVA had itself become a power trust. Lilienthal disagreed vehemently. The TVA, he asserted, was in the business of helping, not gouging, ordinary folks. Liberating, not coercing. True, the idea of an energy "yardstick," determining the cost of producing a kilowatt of electricity, had not exactly worked out. It could not be absolutely proved that federally produced electricity was cheaper than its privately produced counterpart, but Lilienthal liked pointing out that the TVA *had* succeeded in doing two big things: lowering the prevailing rates (in its own and in neighboring regions) and substantially driving up demand for electricity. Before TVA, Lilienthal said, consumption in the valley area was 50 percent less than the U.S. average. After TVA had been in place, the valley average had risen to 25 percent more than the U.S. average.[22]

David Lilienthal was an effective administrator for several reasons. He possessed great energy, was a charismatic speaker, and remained a friend of congressmen and governors even as he remained largely unencumbered with the sort of payback usually required for political support. He thrived in public meetings and was noticeably dedicated to his great enterprise. His enthusiasm, intense to the point of lyricism, showed up vividly in his writing, especially on the topic of electricity and how it transforms land and people. Lilienthal was his own Thoreau, his own Henry Adams. He didn't need poetical help in speaking of the Tennessee River. He is not normally classified as a poet, and yet some of his prose bears comparison, I believe, to some of William Wordsworth's nature verse. Here is Lilienthal describing his own lake district. He talks not of daffodils or passing clouds but of hydroelectric power:

> There is a grand cycle in nature. The lines of those majestic swinging arcs are nowhere more clearly seen than by following the course of electric power in the Tennessee Valley's way of life. Water falls upon a mountain

> slope six thousand feet above the level of the river's mouth. It percolates through the roots and the sub-surface channels, flows in a thousand tiny veins, until it comes together in one stream, then in another, and at last reaches a TVA lake where it is stored behind a dam. Down a huge steel tube it falls, turning a water wheel. Here the water's energy is transformed into electricity, and, moving onward toward the sea, it continues on its course, through ten such lakes, over ten such water wheels. Each time, electric energy is created. That electricity, carried perhaps two hundred miles in a flash of time, heats to incredible temperatures a furnace that transforms inert phosphate ore into a chemical. That phosphatic chemical, put upon his land by a farmer, stirs new life in the land, induces the growth of pastures that capture the inexhaustible power of the sun. Those pastures, born of the energy of phosphate and electricity, feed the energies of animals and men, hold the soil, free the streams of silt, store up water in the soil. Slowly the water returns into the great man-made reservoirs, from which more electricity is generated as more water from the restored land flows on its endless course.[23]

TOP-SECRET GRID

However removed this green valley of the Tennessee might have been from international broils, when the nation went to war in 1941, so did the TVA. Soon a majority of its electricity went for war work. In fact, so much power was needed that hydroelectric generation was no longer enough. The Tennessee River had given all that it could, and so the directors (Lilienthal was chairman by now) built some coal-fired steam-electric plants. In an earlier time critics might have said, "See, the TVA is a power company like any other. We told you so. The TVA doesn't even pretend anymore that the dams are there to help navigation. The dams are there mainly to make power. And now, look, they're adding thermal generators too." But complaining was scarce since this was war and the country needed more electricity.

Lots of aluminum was needed for the war effort, and lots of electricity is what is needed to extract aluminum from the embrace of its rocky ore. Hydroelectricity is what had attracted heavy industry to the Niagara Falls area in Edison's day, and this is what now brought the largest aluminum plant in the world to the Tennessee Valley. Electricity went into the metal, and the metal went straight into the bombers dispatched out over valleys in Germany, such as the Ruhr. How much power was needed? Lots. The amount of electricity used in making a Flying Fortress bomber was equal to the power used in an average American household over about 400 years.[24]

There also came at this time a particular request from the army for more electricity, much more electricity, for a peculiar facility it was building near the town of Oak Ridge, in a remote part of eastern Tennessee. Lilienthal, the man who presided over that hoard of power, was not taken into the confidence of the higher officials as to the plant's purpose even though it was located in his province. The TVA network had always been a sort of sovereign grid within the nation's overall grid, and now here was a grid within the grid within the grid. All Lilienthal knew was that the secretive site constituted an immense energy sink. Every day more and more electricity went in, but nothing ever seemed to come out. What could they be building in there? What kind of strategic metal were they extracting?

Only at the end of the war—actually on August 6, 1945—did the world learn what had been going on at Oak Ridge. To make the needed military electricity, TVA had for years been harvesting gushing water, so-called white coal. When that wasn't enough, TVA resorted to scraping the more traditional black coal from the surrounding Appalachian hillsides to burn in steam-electric plants. Now a new energy-rich material, uranium, a sort of yellow coal, was being put to controlled use for the first time in human history. When separated from its ore and when the specific variety of this element had been selected out through laborious enrichment, a process carried out at Oak Ridge inside the largest building in the world, the net product would be the most precious and dangerous lump of matter in the world. From the bowels of the TVA had come energy needed to stock the device that, delivered by one of those electrically crafted aluminum aircraft, flattened the city of Hiroshima in an instant. The regular war was over, and the nuclear campaign had begun.

COLD WAR GRID

It was time to take stock. If the TVA were a person, it would probably have received a medal for meritorious efforts during wartime. TVA's military work aside, however, one could ask now whether the TVA experiment had been successful. David Lilienthal was unequivocal in saying yes. Of its first job, that of helping to relieve the Depression-era plight of the Tennessee Valley people, living conditions had definitely improved. Literacy, libraries, and soil conservation were up. There had been a huge increase in power consumption. But were the costs worth the investment? Things had improved for valley citizens, but who had paid for this? The taxpayers of the nation. It was said that "the Tennessee River flows through seven states but drains the nation."[25]

Power consumption in the valley had gone up, but consumption just outside the valley had gone up just as much, or even more.[26] Navigation on the river was improved and Knoxville was now a "seaport," linked through locks with the Ohio River. Through its system of dams and locks, the Tennessee was, as Lilienthal liked to say, the most controlled major river in the world.[27] But getting to Knoxville from the ocean was expensive; from the level of the Ohio River at Paducah up to Knoxville was a vertical climb of 515 feet over a distance of 648 miles. The volume of river traffic might not have justified the expense.[28] The steam-electric plants built by the TVA augmented its power to an immense degree, so much so that the hydroelectric component became only a minority of all TVA power produced. Other turns in the tale of TVA: It would eventually become the largest consumer of coal in the nation[29] and, in the course of time, the largest violator of the Clean Air Act.[30]

World War II had seen the defeat of Germany and Japan, but quickly these nations were on their way toward being allies while, conversely, the wartime ally, the Soviet Union, was becoming more of a menace. The bomb technology perfected at Los Alamos and the fuel enriched at Oak Ridge were now at the center of a critical debate over the future use of nuclear energy, a debate explored in more detail in Chapter 8.

Lilienthal, with his extraordinary expertise in operating a large government energy agency, participated in the deliberations about how to handle the new and dangerous form of energy. Partly because of Lilienthal's persuasion, Congress decided that all of the nation's nuclear knowledge, assets, and future missions should be transferred from the army to a new civilian entity, the Atomic Energy Commission, or AEC.

Electrical and nuclear affairs were to be intertwined. Looking over the annals of TVA history, several large epochs present themselves. In the 1930s the TVA was viewed as a tool for declaring war on Depression misery in the Tennessee Valley. In the 1940s the TVA helped fight the war against the Axis powers in Europe and Asia. And on into the 1950s and beyond, the TVA would use a majority of its power to wage the new war of communist containment. The TVA would be an important power supplier for the AEC. It would help deliver the nuclear fuel needed for bombs. The TVA grid had become and would remain partly a military grid.

It was not surprising, then, that David Lilienthal, chairman of the TVA, should become the first chairman of the AEC. The man who had carved out a hydroelectric empire in the river course of the Tennessee would now launch an even larger and more strategic empire spread out in a dozen labs around the country, an empire based on the nuclear activity of uranium and plutonium. He and his associates, and those in a handful of other countries, would constitute a nervous new plutocracy controlling—or at least threatening or preparing to use—the most extravagant form of energy yet devised by the human race.

Electricity had been used to forge the primordial nuclear device, the atom bomb. Conversely, years later a different nuclear device, the reactor, was used to make electricity. This two-way marriage of convenience between nuclear and electric forces, although apparently sanctioned by wartime or economic necessity, has never ceased to cause anxiety. Extraordinary that Lilienthal should in succession have had both these tigers by the tail.

IMPERIAL GRID

Enthroned on a mesh of wire, electricity had triumphed. It had become the paramount utilizable energy. Wind, water, steam, gas, fire, muscle—all had their turn. All were still present at some level, but if you could afford it, electricity was how you toasted your bread and lit your rooms. Electricity had been caught from the sky by Franklin and induced to race through wires by Faraday's invisible force fields. It was standardized by Edison, motorized by Tesla, dispatched for miles by Westinghouse, and domesticated by Insull, whose legionnaires first conquered all competitors and then hunted mercilessly for consumers house to house armed with rates too low to ignore. Electricity had become an affordable convenience. The grid had become an institution. You trusted it. There it was in the lamp next to your pillow and in the wall beside the baby's crib.

The grid enclosed a wild thing—electricity is a form of bottled lightning, after all—which would escape from its wires if it could. But the grid itself, its towers and its machines, shows no signs of being contained. In the 1930s the power network had become large and would get larger, far outpacing Insull's grandest schemes. Electricity would eventually travel to the moon with the astronauts, would enliven computer-controlled robots for assembling Japanese autos in Argentina, and would ever so gently actuate microscopic sensors that deftly tell sick from healthy blood cells in serum samples or effectuate delicate surgical improvements in your eyeball.

A tangible artifact of the diverse grid types can be found in the signature slogans mounted on the sides of the respective powerhouses. In the USSR, Lenin's maxim about communism being the sum of Soviet power and electrification was riveted to the walls and can still be seen at some installations. At the TVA the inscription carved into the dam structure, and still there, reads simply: "Built for the People of the United States." Meanwhile at the Niagara generators, a simple plaque bearing numbers was bolted to the dynamo. On closer scrutiny visitors would see, and can see nowadays at the Smithsonian Museum of American History in Washington, D.C., that the plaque specified legal claim to the generation process being used to make electricity; the numbers were those of the famous Tesla patents sold to, and proudly owned by, the Westinghouse Electrical Corporation.

All these famous names had by now left the scene. Thomas Edison lived on peacefully until 1931, inventing the whole while. On the occasion of Edison's final departure, his old associates inquired respectfully into the possibility of New York City's power being turned off for one minute in tribute. The request was denied.

In 1907 George Westinghouse lost control of his own company, as Edison had lost control of his in 1892. Seven years later Westinghouse died. As for Nikola Tesla, the ever-energetic whirlwind of ideas, he pursued many imaginative schemes—robotics, radio broadcasts, power transmissions through the air, and speculations about neurology—that have earned him a sort of cult following to the present day. He lived until 1941.

Samuel Insull, like Trotsky, had been pursued across the world. He too had been stripped of his posts and scorned by government officials who once were his admirers. In the wake of the 1929 stock market crash and the collapse of the utility conglomerates, Insull had been indicted on numerous state and federal counts alleging fraudulent methods in attracting investments for his holding companies. In a way, Insull's trial was a sort of show trial. He stood as the epitome of all greedy, manipulating robber barons. He was the leader of the Power Trust. His trials received intensive news coverage—celebrity jurisprudence, 1930s style—and his accusers hoped for a long jail term. Insull claimed, basically, that he was in it for the electricity, for the energy, for the adventure of building something really big, and not so much for the money.

Passing judgment, several juries must have believed him because he was acquitted on all charges. He moved to Paris where he lived modestly, but not poorly, in an apartment. One day in 1938 in the Paris Metro his heart stopped. As it must for every man, death had come for Samuel Insull. If one were to fashion a *Citizen Kane* style movie out of this story, the last words to cross the dying man's lips, I like to think, would probably have been "Pearl Street."

In some parts of the world the government—city or national—owned the power grid. In other places, especially in much of the United States, the grid was owned by private companies. The most famous of these, and for many years the biggest, was the one Edison founded at the foot of the Brooklyn Bridge. Owing to its historical role in setting electrical trends and in practically epitomizing the idea of the arch-capitalist company and electricity provider, this utility will occupy our attention now and for some time to come.

THE WORLD OF TOMORROW

Not long after its creation, Edison's old company had undergone many changes; the affiliates went their own ways, the Edison research and manufacturing operations were absorbed into General Electric in 1892, and the part of the business that operated the Manhattan grid was bought by the Consolidated Gas Company in 1899.

It is that grid fragment of the old Edisonian kingdom that will be of interest here. Although it began as a subsidiary of the larger electric company, this first of all the big-city grids never behaved like a vassal. New York Edison was destined always for greatness. From its base in Manhattan the company ran wires along the side of the Brooklyn Bridge in order to hitch itself to Brooklyn Edison. By 1902 the whole of Manhattan and the Bronx were electrified, with the other boroughs close behind. In the 1920s, a period that saw more capital expenditure on electricity than any single decade of spending during the railroad boom of the previous century,[31] New York reclaimed from Chicago the honor of having the world's most powerful steam-electric generators.

In 1936 the combined gas and electric company changed its name to the one it bears to this day: Consolidated Edison. Its motto was "Dig We Must" because its work crews never seemed to stop ripping up some street to repair or expand the wiring. By now they were connected with that other big power center at the diagonally opposite corner of the state, the Niagara-Mohawk Company, via a robust cable kept at 138 kilovolts. Such lines were creating what could be called a United States of Electricity. Its confederated members—local grids in Syracuse, Rochester, Providence, Boston, Buffalo, Hartford—could buy and sell surplus energy in the span of seconds. (Observing this power pool in action will occupy the whole of the next two chapters.) Other such electrical leagues were becoming common, in the United States and elsewhere. A grid of grids, a greater grid, was coming into being.

To see how intergrid cooperation works, look at what happened in 1944 when a fire in the Bronx knocked out two huge generators. A moment before the accident, New York had been exporting 70 megawatts of power *to* the north. One second after the breakdown, 100 megawatts of relief power *from* the north came galloping to the rescue.[32] The New York City grid had stayed intact because of the outside help. Twenty years later, New York returned the favor. On the very day a new tie line was opened between New York and Connecticut, a blackout in Hartford was averted by swiftly switched reinforcement energy from the south.[33] This is one reason why grids are tied to each other.

Con Ed and the electricity industry in general were on the march. Even through the terrible Depression of the 1930s, electricity consumption had mostly been going up. The 1939 world's fair, held in New York, was the ideal place to show off. At the General Electric pavilion, for example, amazed visitors could witness an artificially produced 10-million-volt streak of lightning. For the price of admission, you could see a portable radio the size of a suitcase and be among the first to observe the demonstration of technology that would permit pictures to be broadcast along with sound in one unified radio signal. This early sample of television was powered by good old Jumbo Number 9, the same generator that had played the starring role on that first morning on Pearl Street in 1882. Jumbo had survived the powerhouse fire of 1890, and had been a star at the Columbian Exposition in 1893 and another exhibition in St. Louis in 1904. It retired to Henry Ford's museum in Detroit, where so much other Edisonia was stored. It managed to exert itself in 1932 on the 50th anniversary of Pearl Street. And now here at the New York fair of 1939 it once again compelled electricity to flow through wires.[34]

The world-of-tomorrow theme at the fair was especially prevalent at the Con Ed pavilion, where a gigantic diorama, a stylized rendering of the five boroughs, vividly juxtaposed the past and the future. The City of Light, as the display was called, depicted activity all over, under, and through the metropolis. Electrification, the viewer was reminded, was pervasive in daily life but still charmingly new:

> Yet it is amazing to think that most of the magic we see today—our skyscrapers, our great bridges, even the automobiles and subway trains moving to and fro beneath our rivers, and the great winged ships in the sky overhead—all these man-made miracles have happened in the short span of a human lifetime. For less than sixty years have passed since Edison gave the world its first power station. There, at Pearl Street in 1882, in the shadow of the construction of Brooklyn Bridge, our world of today began.[35]

The fair brochure congratulated readers on their embrace of electricity over the previous six decades. But this was to be nothing compared to the ardor for voltage shown after that. In the period between the world's fair of 1939 and the one in 1964–1965 (also in New York), power consumption in the United States tripled.[36] David Nye, an expert on early American technology, draws a vivid portrait of the tremendous amount of concentrated energy available to Americans courtesy of their 1960s-era machinery. For example, the average home had the use of more energy than a small colonial-era town; a color TV, played for four hours, used more energy than a team of horses rendered in a week; a typical automobile possessed more horsepower than the output of 10 small water-driven mills.[37]

Where does the energy come from? The 1960s looked as if they were going to be an energy-hungry decade. Utility planners must be demographers—more customers means more electricity. On the eve of that second world's fair in New York, Con Ed was preparing for the future with an ambitious slate of new construction, including one of the nation's first privately financed nuclear power plants, to be situated on the Hudson and the world's largest steam-electric plant.

A leading generator manufacturer of the day, Allis Chalmers of Milwaukee, had once built a relatively small 7-megawatt generator (although in 1903 it would have been the world's largest), known affectionately as Little Allis, for Con Ed's East River Station. So when the same company was called on to build an unprecedented big machine for the Ravenswood power station, the newcomer was dubbed Big Allis. For engineers the building of this behemoth was to pass a psychological threshold analogous to breaking the sound barrier.[38] This would be the first steam-electric dynamo to reach the 1,000-megawatt plateau.

Why build something that big? Better economics. In a business where sales are measured in hundreds of millions of dollars, if by choosing one generator design even a few percent could be saved on costs in the construction or operation of a plant, then you would be highly motivated to adopt that design. From Sam Insull's day right up to the 1960s, when Big Allis was built, it was the case that bigger was better: cheaper to build (cost per installed kilowatt of capacity) and cheaper to run (cost per kilowatt of output). Facing the East River just north of the 59th Street Bridge (and within visual distance of several other former world's largest generators—the whole constituting a Pearl Harbor of topline dynamos—Big Allis brought Con Ed's total capacity up to 7,000 megawatts, double what it had been only 10 years earlier.

Just as Jumbo Unit 9 had been the workhorse of Edison's 1882 grid, so Ravenswood Unit 3 was meant to be the flagship of the 1965

Edison grid. More than any Dreadnought battleship, Big Allis was a world unto itself, what with a million feet of cable, 360 miles of tubing, 50,000 electrical connections, 12-story-tall double boilers for making 1,000-degree steam, a smokestack half the height of the Empire State Building, and pumps for forcing 530,000 gallons of water per minute through its condensers. Allis was an energy omnivore, swallowing 1,000 gallons of oil per minute or 2 million tons of coal per year.[39]

Naval powers of the world have usually been worried by rival builders of battleships putting their ships aggressively into the common ocean. So it can be with utilities watching dynamos of overwhelming power being wired up to the common grid. From across the Hudson River, engineers at the Public Service Electric and Gas Company of New Jersey, the main power company just to the west of New York, viewed Big Allis with some trepidation. The fact of interconnectedness meant that in a certain sense generators forming part of the grid next door were part of *your* grid too. A generator from the next jurisdiction could pose a danger to your own grid. Utilities mostly appreciated the usefulness of a working alliance with other power companies, but they were mindful, too, of the price of grid imperialism. Like a miniature sun, a 1,000-megawatt power plant could look like a ball of energy. Flexing its potency, could such a machine burn out another grid's circuits? One New Jersey engineer, seeking extra insurance, argued for installing extra circuit breakers between the systems, just in case.[40]

POWERHOUSE OF NEW YORK

Even more than Texans, New Yorkers like to build big. New York is the Empire State, and New York City is the Big Apple. One reason for this bigness was Robert Moses. To start with, his title of parks commissioner didn't do him justice. Holding a variety of state and city offices for decades, Moses was, after the governor and mayor, the most powerful government figure in the state and maybe one of the most powerful in the country. Moses was practically unfirable; he was to government construction what J. Edgar Hoover had been to crime enforcement.

Moses, like Insull and Lilienthal, possessed an empire in which he could try out vast technological experiments with equally vast social implications. This empire did not spill across state borders—it was centered in and around New York City and other parts of the state—but this was more than enough scope for Moses's ambitious designs. He liked to build bridges, and many of those visible in the city today are his: Throg's Neck, Whitestone, Triborough, and Verrazano. He liked to build auto expressways, dismantling neighborhoods if necessary to do so: Staten Island, Cross Bronx, and the Brooklyn-Queens expressways.

Moses built Lincoln Center, Shea Stadium, and some of the largest apartment complexes in the world. As a member of the state power commission, he helped beef up the hydroelectric resources at Niagara Falls and on the Saint Lawrence River. A hydroelectric plant bears his name. He also had a big hand in constructing the two world's fairs in New York, in 1939 and 1964–1965.

It so happened that one fine day in August 1965 was declared Con Ed day at the fair. Oddly, Moses, the great friend to motorists, did not himself drive a car. He had been delivered to the fair that day by limousine. There he hosted the Con Ed contingent, some 22,000 employees, spouses, and kids. It was a good time for the big New York utility. Just the month before, to great fanfare, the monster generator Big Allis had exerted herself for the first time. New Jersey power officials could have their doubts, but New Yorkers were happy to have the new artificial sun keeping lights on around the city.

With the benefit of hindsight we might say that New York, as a city or as a technological venue or as an engineering powerhouse, might have been nearing the end of some giant ascent. The World Trade Center would be going up downtown, but federal construction dollars would not be so plentiful thereafter, the big building projects would be fewer in number, and there would be a growing sense that the best was behind. Con Ed's biggest construction days were also behind it, although this melancholy fact would not have been known to those assembled at the fair for Con Ed Day.

Moses's career was coming to an end. The world's fair was a bit of a disappointment. The 1964–1965 fair, attended by tens of millions, was hardly a fizzle, but by some accounts it did not quite match the magic of the 1939 fair, especially when it came to the allure of electricity. One sociologist put it this way: "1939 had been the promise, 1964 the fulfillment." "By 1964 the future had already happened."[41] Take television for example. In 1939 television (radio with pictures) was a dazzling expectation. By 1965 television had not only arrived but was highly embedded in the visual landscape. Other wonders—electronic brains, automatic figuring machines—were now performing thousands of times faster than the smartest mathematician. In the oceans, voyagers had visited the seafloor. In the sky above, men had circled the globe in 90 minutes and had walked in outer space tethered to a frail craft. The moon was next.

Why did Con Ed deserve a whole day to itself at the world's fair? Because it was Con Ed. With 26,000 workers, it was the largest business-managed utility in the country. Con Ed had 180,000 stock holders and was the biggest real-estate taxpayer in the city.[42] If you went to work for Con Ed, it was not uncommon to work there the rest of your life. It was not unusual that your father, grandfather, and maybe some aunts, cousins, sisters, and daughters worked there too.

The company was a small civilization by itself. It had its own softball leagues, an Emerald Society for Irish-American employees, generous widow benefits, and a bureau for veterans' affairs. It had clubs for blood bank gallon donors and regular banquets for 25- and 40-year staffers. Athletic and culture activities? There were clubs for angling, bowling, archery, bridge, skiing, chess, photography, organ playing, stamp collecting, and many other pastimes. There was a Con Ed Society in Florida. Safety records were proclaimed in the company newsletter, along with new babies, deaths, baking awards, civic volunteerism, and scouting merit badges. This all-encompassing company had earned special treatment.

On Con Ed Day, Robert Moses, president of the fair and a man who would have been one of Con Ed's few historical rivals in shaping the physical landscape and environment of New York City, made a gracious speech. He welcomed the utility's employees with timely flattery: "Consolidated Edison Company, which heats, lights, and powers our city. Without you, we would be paralyzed and our future would be dark indeed."[43] This proposition would soon be put to the test.

5

WORST DAY IN GRID HISTORY

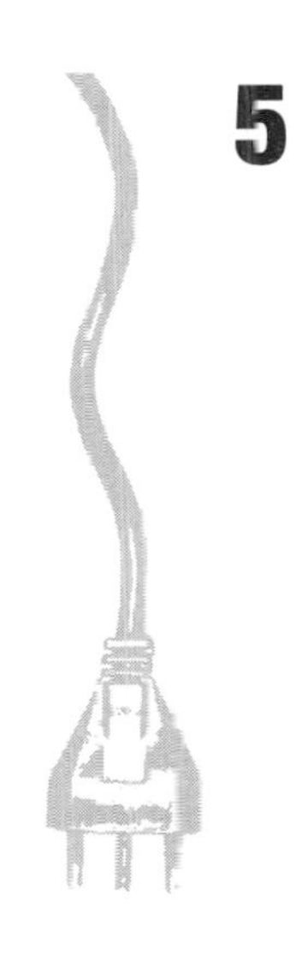

Edwin J. Nellis possessed what any person in his position could want. He had more wattage under his control than all the feudal kings of history. With a flick of his wrist he could direct a river of energy more forceful than that pent in a million-horse stampede. He could command more light than might be massed in the collective candles of a thousand cathedrals. He was neither potentate nor pontiff but the chief power dispatcher on duty at Consolidated Edison's Energy Control Center on Manhattan's Upper West Side.

The setting could almost be called biblical. Although it doesn't look like a throne, the system operator's chair is a seat of power greater than that of the storied rulers of Sumeria. Only now the Mesopotamia in question, the land between the rivers, lies not athwart the Tigris and Euphrates but between the Hudson and East rivers. The man in the main chair dispenses energy to the inhabitants of that land of 9 million inhabitants.

The terms of this rulership are anything but reverential or even respectful. No, the fealty is strictly mercantile. It is an oligarchy of seller and buyers, with prices set not by barter or haggle but by regulatory statute. On the one side is the utility company, in existence to clear a profit. On the other side are the multitudes, 3 million billable households, eager to accept the energy but resentful of the per-kilowatt-hour rate, the highest in the land, and skeptical about the incessant ripping up of roads needed for underground repairs.

We are a corporation, the company vice president would declare politely. Although service is paramount, dividends do have to be paid. Nothing wrong with that, is there? Like any other large business, we must cover expenses and produce a return on investment for those who lend us the capital used to build those massive generators and all the other modern equipment installed on the grid in a process of continual upgrade and improvement. If there were no profit, there would be no investment, no new technology, no new capacity to meet the increase in demand. We are pleased and privileged to deliver electricity, but it has to be paid for.

The company engineers are intensely pragmatic. They think about the customers' high-torque drills, their hedge clippers, their window fans yearning to turn free, as being an aggregate load on the grid. It is, after all, the engineers who designed the grid and who keep designing, since the grid gets ever larger, never smaller. Like a 14-year-old boy, the grid seems to outgrow its clothes almost every day. Consequently, the grid must be retailored perpetually.

Why is supplying electricity on a grid not like supplying bread from a truck? If a baker is short a few loaves, he apologizes and tells the customer to return earlier tomorrow. A few loaves too many and the baker can make breadcrumbs with the leftover. But with electricity there is no leftover. It has to go somewhere all the time and can't lie around unused on a shelf at the back of the store. At all times the generation must meet the load. Supply and demand are exquisitely squared off.

The delivery not only has to be economical and reliable but also timely and exact. When a room light comes on here or a motor starts there, this represents a sudden additional load on the system that must be answered immediately by a proper measure of watts from the wall socket. Electricity supply is very hand-in-glove. Every step forward that Fred Astaire takes must be reciprocated by Ginger Rogers going backward, in heels.

The grid has inbuilt redundancy. If one generator or cable goes bad, a second or a third should automatically shoulder the extra burden. But in an elaborately complex system—one made from lots of swiftly interacting components—it doesn't always work that way. The linemen most particularly concerned with the continuity of service, the ones who wear hard hats and drive those tool-crammed vans, do daily combat with a variety of maladies: cracking insulation, expanding joints between fitted parts, frayed materials, settling ground, tarnished contacts, leaking water. All of these unwanted departures from the norm can lead to criticality. The system can work itself into a fragile condition in which even a tiny problem can quickly escalate into a large problem. The grid is not inert. It does things and things are done to it. The grid can be an amplifier of trouble.

Indeed, today's saga is about trouble. The watchwords will not be profit and loss but chaos and darkness. The first sign that something was wrong came not from the high-tech console readouts, not from the computers. No, it was the fact that the lights in the room were dimming; in the very room where the electricity for 9 million people was controlled, the lighting was uncontrollable.

A moment before, Edwin Nellis had been congratulating his colleagues on their having successfully matched generated power with customer load. Millions of residents were just then maneuvering through the high moment of that day's evening rush hour. Many were still riding in numerous planes, trains, and automobiles. The cusp of peak demand for the day had already been reached. Con Ed's home team of generators was working normally. Furthermore, although the city grid had more than enough resources of its own, it was importing a modest amount of energy from the friendly allied utilities up the Hudson and elsewhere, taking advantage of cheap upstate hydroelectric power. So far so good.

But then dials on the control board started acting strangely. What had been a net import of power abruptly flipped, in the course of a minute of so, into a net export. Energy was being sucked out of New York City and at an increasing rate. Some strange eruption on the grid was happening to the north. The dials swung rather more to the right now. This was no mere flicker of recognition. A rather loud alarm was sounding, obliging Mr. Nellis to reach for his telephone.

OUT TO SEA

The electrical grid in 1965 was not so different from what we have now. In its physical look, in the strength of its dynamos, or the high voltage used, it's essentially the contemporary grid. In any case, the grid's enormity is too big to fit into your head. You can't encompass it. The thousands of miles of overhead heavy-duty transmission lines and the many more miles of smaller wires spooling out underground and through walls are too numerous to count and too spread out in area to be grasped easily. Nevertheless, we will now attempt to wrap our minds around the Eastern Interconnection, which is but a portion of what counts as the North American grid. Even this portion is dauntingly huge, taking in much of southeastern Canada and the northeastern United States.

To set the stage for the exciting event about to take place on November 9, 1965, please picture the extended grid as being a vast internal sea of energy. If Genoa, Tunis, Barcelona, and Beirut all look out onto the Mediterranean Sea and draw fish from it, then so, in an important sense, do Rochester, Boston, Toronto, and Providence all look out onto a common voltage seaway and draw energy from *it.* A curious and useful property of electrical circuits is that power made in Queens, New York, can be ladled out of this sea a fractional second later in Hartford, Connecticut. Even if landlocked as reckoned by geography, all these cities of the grid are coastal as reckoned by electrogeography.

This energy ocean can't be seen, but it's there. It edges right up to your home. Waves are lapping 60 times per second and not just on some shore out back but in the fabric of your walls. They wash up and fill the interstices of all habitable places. This energy sea sends waves up every single cord. Plug in another machine in the corner of the room and instantly the energy sea extends in that new direction too.

Instead of a universal sea level, the internal electrical sea is maintained at universal levels of frequency and voltage. Not only do the waves arrive everywhere at the same rate—60 times per second in the United States, 50 times per second in France—but in the same phase with the same choreographed movement. That is, the waves are totally synchronous. When the electric force is swelling to a maximum inside the wires running through a stoplight in Orange, New Jersey, a coordinated swelling of force occurs in television sets in Hamilton, Ontario.

Despite the rich diversity of electrical commerce around the internal energy sea, all aspects of power movement—whether into or out of the sea—are neatly timed, adjusted, and transformed with pinpoint control, so that when Niagara makes and Syracuse takes, all the electricity that's needed gets sent and all the electricity that's made gets used. What other area of civic life, carried out across a field of play so oceanically large, delivers a product in huge volumes a microsecond after receiving the request? It works really well and according to plan more than 99.9 percent of the time.

That means, conversely, that for a few minutes or hours per year things *don't* go as planned. Actually, the span of time concerning us presently will be about 12 minutes long.

ON THE ENERGY FRONTIER

On November 9, 1965, 10 people died in a stampede on a rail bridge in India. In Vietnam hundreds were killed in jungle battles between Vietcong forces and the American army. In the United States, former president Dwight Eisenhower had a heart attack and a man set himself on fire in front of the United Nations headquarters, where diplomats were debating whether the mainland Chinese delegation should be seated.

These events all appeared in the newspapers but were even more vivid when they appeared on television, the electrical machine that has done so much to sculpt popular culture and fill up free time. The human mind craves entertainment, and television is only too glad to oblige. Air conditioning coddles our outsides, but television tickles our insides. Air conditioning merely brushes the outer dermal layer, while a TV program splashes imagery against the retina at the back of the eye, whence it moves nervously along to the visual cortex at the back of the brain. If electricity were a religion, the televisual glowing box would qualify as the most venerated object, the totemic device before which more people spend more worshipful hours than any other idol.

A favorite show can be a regularizing influence in one's routine.

Coming at the same time on the same night of the week, the show induces anticipatory pleasure. It becomes practically addictive. Many plan an evening's schedule around the airing. What happens if, say, an electrical disturbance intrudes? (Did the lights flicker just now?) Some might experience a syndrome of withdrawal symptoms—irritability, a vague feeling of letdown, even desperation. Other electrical gizmos aren't nearly as ingratiating or potentially debilitating.

Recall that "grid" is deliberately being defined loosely; the grid, in my usage, doesn't necessarily stop at the wall socket but continues right into the appliance. The grid is not just the wire or the utility making the current humming through the wire. It's also the *consumption* of the energy and maybe, to push the concept of grid to the edge of usefulness, the *functionality* of the appliance as well.

We now encode big parts of our national mythos in waves, the radio waves shooting through the air or the infrared waves arriving by optical fiber. To learn about the eternal truths, the ancient Greeks made pilgrimages to the oracle at Delphi, where portentous vapors poured out of a crack in the earth. Nowadays we visit the oracle of television, which displays portentous signals on a glowing screen.

The most popular program in November 1965 belonged to a genre known as the Western. Future sociologists might be surprised that modern city folks, marooned on their asphalt islands, should be so interested in cattle drives, shootouts, and the Caucasian conquest of trans-Mississippi North America. Why inhabitants of high-rise apartments in the Bronx should be obsessed with the wizened inhabitants of low-rise barns in Wyoming cannot be answered in a few pages. The answer might have something to do with the willful act of conquering or "civilizing" nature. And just as wagoners and the cavalry were carrying this mission into the geographical American West, so electricity had, decades later, performed retroactively the same role in civilizing (bringing the benefit of powered machines) to the Technological West—that is, to all those places still unfortunate enough not to have regular utility service.

As we wait for the start of the sequence of events that made November 9, 1965, perhaps the most significant day in grid history since the debut of Edison's grid four score and three years before, why don't we return to yesteryear to re-create the West from the electron's point of view. Therefore, please picture in your imagination the radio-wave encrypted version of *Wagon Train* or *Bonanza* or *Rawhide* reaching your rooftop aerial, where electrons, perfectly preserving the blueprint for depicting a herd of cattle crossing the wide prairie, must make their way from aerial to receiver. Surviving the rigors of a tuning circuit, where competing broadcast programs are nulled out of existence, the desired signal is amplified by the intervention of fresh electrons from the TV's power supply, which siphons sustenance directly from the energy sea. To portray wagons forging across the Colorado Plateau in a snowstorm, a microscopic trek inside your set must take place—herds of electrons are guided along their own cattle drive across the electronic equivalent of mountain passes, river fords, and desert plains.

The electron batches for reembodying the western saga or, for that matter, the weather report or a magician on a variety show or a production of *Hamlet,* are sent to a set of electrodes that control a concentrated charged flow coming from the back of the set toward the front. This mechanism, the electron gun, shoots the electrons in the direction of the picture screen, the business end of any TV. In effect, what you had in your living room was a small-scale particle accelerator. The projection of images would later rely on solid-state technology, but in 1965 bulky picture tubes were necessary.

(And at just this moment there seems to be something wrong with the set. Somebody should get up and make an adjustment. Wait, the thing just fixed itself. The program continues.)

Our total attention is fixed at the screen, where the cattle drive is rendered as pinpricks of light exactly where the electrons excite phosphor molecules embedded in the screen. Bits of bright or dark are systematically splayed out in thin stripes. A new set of stripes, each a necklace of shaded tones, is refreshed 30 or 60 times per second across the screen, giving the eye the impression of vivid presence and continuous motion. The light moves through your eye, eventually striking the retina, a sort of biological antenna. Your brain then does the rest of the work.

In summation, symbolic images of make-believe characters representing a reconstructed West are transcribed into a carefully crafted light show flung onto a screen. The energy necessary for maintaining the illusion from instant to instant is provided by a series of electrical waves rippling through an internal electrical sea filling wires from Ontario to Virginia. The projected images are in turn transcribed into neuronal patterns inside your head. This elaborate powered phantasmagoria constitutes, in my opinion, the electrical grid's most profound and subtle task: stocking your neocortex with vicarious experience.

Miles away, maybe hundreds of miles away, something happens to interrupt this marvelous sequence. The last radio signals arrive at the rooftop antenna. The last signals are amplified, a final burst of electrons is aimed at the phosphor screen, a partial streak of pixels lights up on the screen, giving the viewer a fractional view of a cowboy on a horse. Inexplicably the complete cowboy fails to make his scheduled appearance one-thirtieth of a second later. The cattle drive ends, the West evaporates, the set goes blank, and we're left to figure out what higher powers are at work here.

GRID WEATHER

It is 5:16 p.m. on November 9 and in the next 12 minutes the largest electrical blackout the world has yet seen will unfold, but the participants in places like Hamilton or Elmira or Pawtucket don't yet know this. To people around the shore of the great electric sea, the disturbance at first will seem to be something that is happening to them alone. By looking out the window they will notice, perhaps with relief, that it has also affected their neighbors, and maybe their town, possibly the whole state. Later, from the radio, if they have access to a working radio, the wider implications finally start to register. The physical fallout of electrical shutdown can spread in seconds. The social fallout would take more time.

On November 9 the storm that strikes the Eastern Interconnection is not some external storm gathering strength in the tropics and moving north. Rather it is an internal storm of indeterminate origin. In the first minutes even the size of the storm is not known because it exists only on and in metal pathways hidden from human vision. Wires for hundreds of miles around know where the problem is, but *they're* not talking, except to each other.

The city of Rochester, famed for its snow and its music school, often imports power from Niagara. On this day Rochester Gas & Electric (RG&E), the local utility, is buying 200 megawatts from the supergrid. In other words, Rochester is taking more energy out of the sea than it is putting in, the difference being made up by places like Niagara, which has a surplus of hydroelectric power.

Suddenly at 5:17 p.m. Rochester's imported electricity just stops coming. With no warning, it vanishes entirely and along with it the ability of the local grid to deliver energy. Niagara authorities would later declare that RG&E had "stopped taking" the power. Au contraire, RG&E would argue, Niagara had stopped sending it.[1] Much of the city is merged with the gloom of oncoming evening.

What is happening? No one yet knows. Rochester is not to be alone in its deprivation. Further along the coast of the energy sea is Syracuse, and it now turns black. If you were a resident of Syracuse, you too might think you were being singled out for rough treatment, but you'd be wrong. The erosion of electrical service is accelerating. Why should Utica be spared? It goes like this on through a thick tier of population across the waistline of the Empire State. The bottled lightning normally coursing safely through the natural gates and alleys of the grid is at this moment seriously misbehaving.

In some undefinable way the electric sea has roused itself. This rebellious act does not go unnoticed by sensors tirelessly keeping watch. They see the squall coming well enough. Designed to sniff out irregularities in voltage (the pressure behind the electricity), current (how much electricity is flowing), frequency (the repeat time of the electricity), or phase (how far along into the present electricity wave), the faithful auditors of the grid are hectically at work. Their microbookkeeping is reported not quarterly or monthly but secondly, and the findings are not good. The energy sea is turbulent, and small-craft warnings are being issued.

In some cases the warnings are heeded automatically. For instance, the sentinel guarding the Pennsylvania–New York border spots the trouble to the north, determines that the state of affairs has swayed an intolerable amount outside prescribed levels, and triggers the opening of a circuit breaker. After a convulsive retraction of metal contacts—interrupting current humming at hundreds of kilovolts is a process that can be accompanied by a spark several feet long—the local hookup between the Pennsylvania–New Jersey–Maryland (PJM) system (a power pool within the larger pool) severs itself from the Eastern Interconnection. No human intervenes in this action. One machine talks with another, and in a consultation lasting less than a second, a decision affecting millions is made. The breaker is told to open and it does. Pennsylvania has put up its hurricane shutters.

O CURSED FATE

Things are different at Con Edison's Energy Control Center. Clearly something is rotten in the state of New York. Instruments know there is a storm on the energy sea, but no automatic action has yet been triggered. The decision to sever the city from the regional grid will be in human hands. On November 9 those hands belong to Edwin Nellis, a dedicated Con Ed man for 41 years. In the Shakespearean drama about to unfold, Mr. Nellis will play the part of Hamlet, a central character overtaken by events not of his making, a man whose conscience is stricken by conflicting forces and obligations.

Quickly now, what do we know? First, there was a surge of power *into* the New York City network and not just from the north but also from New Jersey. This surge quickly tripped circuit breakers, isolating New Jersey, parts of Brooklyn, and Staten Island from the rest of the boroughs. So, already some of the city is no longer under Edison control. But the greater danger lies to the north. Nellis is on the phone with Syracuse, where they know nothing. They're already sitting in the dark. Rockland and Orange counties are still up and running, but they're desperate. Send us energy, they plead. An operator up the Hudson, near West Point, has worried himself to a decision point; he cuts his sector of the grid loose from the rest.

In the coming minutes, the problem for Con Ed will be not one of surplus but of shortfall. There is now, you can tell by the dials, a huge *withdrawal* of power up the Hudson. The electricity balance is bad north by northwest. The man in charge has to decide what to do. The controller might address the problem by shedding some of his load—cutting off power to specific neighborhoods, a thing he does not like to do—or quickly increase the generation of electricity, or both. He must be cruel in order to be kind. He prepares to shed load. As for more power, this has already been requested. The Con Ed machinery is taking action against the sea of troubles—the furnaces are banked and the steam is rising—but can the generators make the extra electricity soon enough? In this time of trouble, Big Allis is being asked to produce more, and she is rising to the challenge.

Things were going so well. Rush hour had been smooth. Now, energy is deserting New York. Problems are arriving not singly but in battalions. Why is the system coming down in so many places? It isn't the weather. There has been no telltale thunder, no fall of snow to weigh down the power lines, no reports of failure at any major station. With weeks or months of coal piled up near the boilers, no responsible powerhouse could have run out of fuel. There has been no foreign incursion, at least none that we know of.

Edwin Nellis could be forgiven for asking himself why all this is happening to him. Why today? Why on his shift? Here's what Hamlet said, trying to calm himself, when he too was in a tight situation:

> There is providence in the fall of a sparrow. If it be now, 'tis not to come; if it be not to come, it will be now; if it be not now, yet it will come. The readiness is all.

On November 9, however, Nellis is not ready. In the passing of a few minutes he has become less like an omnipotent king and more like the Wizard of Oz, an ineffectual turner of knobs behind a curtain.

Could the gods of electricity, who had taken pity on mankind and who had entered into a firm covenant with them, be revoking their blessing? Edison and his tribe down through several generations had been in awe of electrical power. But the old fervor was gone. Although mortals still tendered their burnt offerings in the form of broiled coal, they now seemed to take for granted the potent gift of high voltage. It was no longer special. So perhaps the gods had sent a vivid sign. And there was, lo, a great departure of energy throughout the land. If you were apocalyptically inclined, you might say at this moment that the cities of the plain were being smote.

Toronto had been one of the first to be cheated of its energy. Subways, streetcars, and trolleys lost their means of moving, and airports switched to auxiliary power. With the energy gone, molten copper re-solidified in the furnaces of the Wolverine Tube Company. At Christie's Bread Company, where conveyor belts slugged to a halt, 2,000 loaves were stranded in the middle of the oven, where they solidified in a horrible way.[2]

The primary utility for this area, Ontario Hydro, is mystified as to the cause, although preliminary indications show that the problem lies to the southwest, near Niagara Falls, probably on the U.S. side of the border, or else farther off, near Rochester. On the other hand, a confusing blast of energy has just whipped counterclockwise around the *eastern* end of Lake Ontario. How do you explain *that* if it were Rochester's fault? Neither country wants to be proved responsible. No utility wants to be the goat.

Grid weather is unseasonably mixed. Gusts of energy show up in unexpected places, followed by huge withdrawals of energy. "God was very good to us," said Sister Elizabeth Ann, at the Our Lady of Victory Hospital in Buffalo. "We didn't have anybody in crucial areas that needed electrical equipment. There were no surgical emergencies or maternal delivery cases." Buffalo, if you will recall from chapter 2, had been a mecca for big business precisely because of its proximity to Niagara Falls and its cheap hydroelectric power. So when all that Westinghousian alternating-current power was withdrawn, when all those motors operating in the Tesla manner were bereft of their missions, there was bound to be industrial trouble.

If you want to know where millions and billions of dollars in business losses come from, start with Buffalo. At Chevrolet the assembly of automobiles halts as if someone had flipped a switch. At Du Pont several batches of chemicals are spoiled at great cost. At Dunlop Tire and Rubber Corporation 1700 tires, curing in molds, are wastefully lost. At the Worthington Corporation great damage occurs at those surfaces where powered blades in drill tools and milling cutters dig into resistant materials. At Republic Steel some water-cooled doors for the big hearths are ruined. At Bethlehem Steel a voltage fade-out automatically trips off the production process. Three employees of Hooker Chemical Corporation are overcome by escaping phosgene gas.[3]

The misbehaving energy sea makes quick work of Albany. The governor of New York himself is gone, as well as the lieutenant governor, so a lesser official steps in as emergency coordinator. Where does he get *his* news? From a transistorized radio tuned to a rock-and-roll station still on the air. Around Albany the blackout has differing effects. In the Dolliwog Lounge at the Ten Eyck Hotel, for example, Miss Suezenne Fordham braves the dark and goes on with her piano artistry. At the capitol, Jackie Robinson, famous for breaking baseball's racial barrier, cannot overcome the electrical barrier, and his civil rights press conference has to be cancelled.[4]

The cities situated around the internal energy sea are widely spaced in the geographic sense. The interstate highway between Buffalo and Boston takes hours to traverse. In the electrical sense, however, that same distance can be covered in a fraction of a second. In terms of copper wire, Boston and Buffalo practically coincide. A power drill in Ottawa, tethered to the universal grid by an extension cord, and a dehumidifier in Amherst are scarcely a heartbeat removed from each other in terms of the grid's own secret time frame. Two very different perspectives are at play: The intercostal cities see the grid as a go-between, an extended *extrinsic* thing, a servitor bringing sustenance over wires. Complementarily, the grid would see itself—if it were a sentient being—as a centered *intrinsic* thing, a gathering and spreading of energy. It would regard the cities as being just so many loads, mere bunting hanging pendently from the grid's banisters.

For the beleaguered dispatchers of grid energy, this gusty squall of confused currents has come so suddenly that there is no time to stand back and look for clues or patterns. There is, however, one great observation that can be made. The problem mainly to the west of the Hudson had been the sudden presence of *too much* power. The surge quickly bore down on the utilities in its way, triggering relay mechanisms that, operating according to correct procedure, protected the rest of the transmission line downstream and other valuable assets such as generators and transformers from harm by opening a gap in the line.

East of the Hudson the problem is now more one of *too little* power. The main tide of electric supply across New York from Niagara to the east has broken down somewhere between Rochester and Syracuse. And once this normally beneficent east–west Silk Road of electricity is cut off, New England to the east and Gotham to the south are left in the lurch.

THE LAST BEST HOPE

In this province of the Eastern Interconnection, the nearest thing to a Niagara-sized concentration of energy is Consolidated Edison's colossus generator, Big Allis. Before the storm blew up, Allis had been coasting, spinning out only about half the power she could actually muster. Woken to the spreading peril, Allis is now cranking up to full potential. If the electrical crisis were told as a children's story, Allis might be the little train engine that attempted to climb the mountain and rescue the boys and girls on the other side: "I think I can . . . I think I can . . . I think I can."

In the Con Ed system as a whole, more than 1,000 megawatts is held back for emergencies, a "spinning reserve" in the form of extra output that can be summoned from generators already operating. The New York grid is one of the mightiest in the world and can furnish all the energy needed for its own grid, but like many other utilities it avails itself of the cheaper electricity made available elsewhere in the pool wherever it can. To be exact, it was importing 360 megawatts from the Niagara/Mohawk Power Corporation when the crisis began. Now, in a matter of only a few minutes—hundreds of seconds—the whole flow has turned around. A 360-megawatt deposit has become a 750-megawatt withdrawal. Connecticut is drawing another 250. With this kind of run on the bank, the vault soon would be empty. They say it is better to give than to receive, but this is too much. Big Allis, working hard to save the city, has already brought up 100 extra megawatts. If only there were more time.

The 1965 blackout is turning out to be a harbinger of the 2003 blackout. The Eastern Interconnection, what's left of it, is foundering. The levees are weakening. As the power spreads too thinly, the frequency starts to drop. This effect is equivalent to an organism losing blood pressure. When the frequency falls too far, the pumps serving the big generators turn themselves off, and this in turn cuts off the generator itself. As you can plainly see, this only deepens the crisis. Too much and you faint. Death by underfrequency.

Edwin Nellis and his confederates race from dial to dial. They turn cranks and make calls. What is their next course of action? Shouldn't they desert the decaying grid and at least preserve their city against general ruination? But what about their pledge of assistance to those in need? Only three minutes earlier, Mr. Nellis's life had been so simple. In that short interval the opportunity for greatness had been thrust upon him. To cut loose from the grid or not to cut loose—that is the question.

At 5:21 p.m. the blackout has come for Boston. Electricity itself is ephemeral, so the *absence* of electricity should be an even bigger nothingness, but it's not. A blackout, as it settles over a large city, is a *something.* It's like wet snow. It has weight. It's a form of anti-energy, and in Boston it's everywhere. Immediately there is the same paralysis: stoplights, gas pumps, conveyor belts. In the voltage headquarters for the city, at Boston Edison, 18 people are stranded in one elevator alone. The Massachusetts National Guard issues its first general alert since World War II. In nearby Chelsea, officers attending the annual police benefit banquet pin their badges right over tuxedo lapels, bid their wives adieu, and head out into the chill night to face the situation.

Reactions to a dramatic shift in one's environment can bring out the best and worst in people. The sudden loss of power all over the Northeast is being met, in these first minutes, with a combination of resignation, dread, and resourcefulness. Mostly the fabric of civic life is holding together. Aren't New Englanders supposed to be tough, stoic? And so it is. People are generally friendly, many jovial. Some direct traffic.

At the state prison in Walpole, Massachusetts, however, the prevailing sentiment is rage. Rage at jail confinement, bad food, mean guards—the exact spark is not yet identifiable. What is known is that the inmates are furious and that their stifled emotions are being vented against any furniture or exposed plumbing within reach. These desperate men, residents of the maximum security area, are looking for hostages to take. Tear gas is fired by retreating guards, and reinforcements have been summoned.[5]

Across the state line, Rhode Island has lost its power and its government. The top man, the governor, is out somewhere over the Pacific Ocean returning from a visit to Vietnam. The lieutenant governor is in Boston, presumably in the dark with everyone else. Number 3 in the order of succession is the president pro-tem of the state senate, and he happens to be in Puerto Rico. The fourth man, the secretary of state, is in Hawaii. Number 5, the first deputy secretary of state, is actually in Rhode Island, but doesn't know he's the acting chief, so it falls to the governor's executive secretary to get emergency procedures started.[6]

For Connecticut the blackout descending in all directions will be the first interruption of power since the huge hurricane of 1938. Now in 1965, Hartford's grid is moving rapidly toward the same critical tipping point of no return that has obliterated circuits to the north and west. But wait. Here there is a bright spot. A control operator, with his hand on the switch, is about to make his decision. Automatic cut-out is seconds away, but at the last moment the operator acts. He throws the switch and ruptures the ties to the power pool. This has the practical effect of closing the fortress gate to the city and pulling up the drawbridge. His action preserves electricity in much of Hartford, which remains an island of light in a sea of dark.[7]

Down in New York City something else happens.

THE REST IS SILENCE

Edwin Nellis, struggling against a sea of troubles, can't decide what to do. He works for Consolidated Edison. His duty is to the shareholders and the company's customers. Millions, although they do not know it, are trusting to his judgment. And what does *he* trust? He's been a company man since the 1920s. He trusts the guidelines, which clearly suggest that a utility company should help its allies in time of need. And he has seen to that. Meetings of officials from the various utilities and frequent phone calls among the participating operators all testify to the unified nature of the grid fraternity. They talk with each other every day. You can't just throw that away because of some troubles to the north. Hartford or Orange County might cut themselves off from the pool, but this is New York City we're talking about now. Con Ed, the mightiest powerhouse anywhere, doesn't just cut and run.

News of Long Island's defection from the interlink comes at 5:23 p.m., not that Long Island customers will be spared from disaster. Minutes later their grid crashes too. At Brookhaven National Laboratory, 60 miles out on Long Island, the power goes out at the research nuclear reactor. Its control rods, exactly as they are supposed to do in case of emergency, automatically slide down into place among the fuel elements, cutting off interactions inside the reactor's core.[8]

As the moments zip past, still more power is being squeezed from Big Allis and her allied generators. These machines are practically busting their rivets and still it isn't enough. Shedding load, long a procedure of last resort, is now seen as an absolute necessity. Sacrifice some specific neighborhoods? So be it. Off goes Yorkville, a posh part of eastern Manhattan. Not enough? The West Bronx gets sacrificed. East Brooklyn is axed.[9] How could it have come to this?

After blackouts in 1959 and 1961, Con Ed buffered itself with a series of nimble measures to protect the New York grid from future failures. All these "improvements"—the new computer-automated control room, the new tie-lines to the other grids, Big Allis herself—are proving to be a Maginot Line. The outer edge of the Eastern Interconnection is in retreat on every front. Edwin Nellis is trying to stand up against this deluge. Does he have the proper heroic Hamlet-like emotions—pluck, courage, and perhaps the fatalistic feeling of being crushed by some malicious outside force—or is he just overmatched and confused?

Help the other grid, the book says. But another guideline in the operator's manual suggests, sensibly enough, that if an overload or power drain on the interconnected grid persists, putting your own grid in danger, the operator is justified in cutting free of the problem. Here is the central dilemma. How can you reconcile these two recommended actions—cutting free and holding firm? Which guideline was more important—self-protection or standing by your friends? Edwin Nellis would soon find out.

The storm is not abating. The quality of the energy—volts and cycles per second—is now suffering and soon the machines that make the electricity would themselves start failing, all within seconds. Time, the passing of one moment after another, is an eerie thing because you can't grab hold of it. Even millionaires can't buy more of it. You can measure elapsed time but not time itself. Two different system operators make their respective decisions, two seconds on either side of an invisible blink of time. The consequences will prove to be very different. On one side of the blink, the gesture works. The operator in Hartford throws a switch, sequestering his circuits. Hartford is saved.

In New York the operator throws a switch, but it's two seconds too late. Edwin Nellis, the logbook will show, gave the order to secede from the union. He made the decision to break Con Ed free from the albatross around its neck, a process requiring the pushing of eight separate buttons in the Manhattan control room, which in turn activated giant circuit breakers at the Con Ed substation in Duchess County.[10] But it all comes too late. The trip-offs, marking interruptions along myriad circuits, have already begun. They resound through the Con Ed system like uncelebratory champagne corks popping. The murderer's row of mammoth generators ranked along the East River were, just a moment before, capable of more power production than several Niagaras, which is to say they could liberate more energy in a day by frying West Virginia coal than could be extracted upstate from the gathered waters of the entire Great Lakes watershed. Now nothing.

A long time ago, at the height of the Ice Age, so much of the world's water was invested in northern glaciers that many other aqueous preserves were left destitute. Consequently, the Mediterranean Sea came to be an empty dustbowl.[11] And so it was, eons later, at that quiet moment on November 9, that the internal electrical Mediterranean—my fanciful name for the Eastern Interconnection—came to be empty as well. With the exception of some islands here and there, no electricity flowed between Syracuse and Montpelier or from Plattsburgh to Providence. No electric waves rose and fell in the manifold sea of copper wire.

The energy that 12 minutes earlier had coursed among the allied cities had retreated backward into machines already starting to cool off. By just such an amount, a like measure of coal went unburned in furnaces. Furthermore, trillions of uranium-235 nuclei did not fission at the cores of reactors, and millions of gallons of water at Niagara plunged down to the level of Lake Ontario without producing electricity. Power had been snatched away from millions of customers, including some of the richest and poorest citizens in the land. Electricity is a great leveler.

This was to be the worst day in grid history. Worse than August 14, 2003? In terms of the shock generated, the loss of confidence, the changes set in motion, and because it would later nostalgically be seen as marking the passing of a golden age, yes, 1965 was worse than 2003.

When the wall clocks stop in New York on November 9, it is 5:28 p.m. For the first time in the history of Consolidated Edison, nearly the entire system thuds to a halt. Allis's rotor has so much inertia that it will continue to spin for more than an hour on its own. And this spinning, this gradual coming to rest, will be the ultimate end to the historic blackout of 1965. A dovetail of technological collapse, taking its full course, will grind to a halt here in the middle of this billion-watt energy machine in Queens, New York.

Funny, but no one had contemplated the possibility of there being *no electricity* in New York City's wires anywhere, and so even as the massive central shaft at the heart of Big Allis, the largest single machined part in the world, is gradually revolving slower, slower, slower, there is no power for the oil pumps that should be lubricating the many places where metal rubs against metal. Consequently, the fretful final moments will not be so smooth. Allis's bearings, deprived of their lubrication, are starting to melt.

6

THIRTY MILLION POWERLESS

Go back 12 minutes. Imagine that you have just flown up the eastern coast of the United States. In the past half hour alone the captain will have invited you to gander to the left, where you saw in succession the lights of Washington, Baltimore, and Philadelphia. And then along comes the jewel in this crown of cities, Gotham itself, the most illuminated place on the planet. With buildings all but invisible, the conjoined street lamplight of New York City looks molten, like some natural wonder of the world.

Here is what happens in that airspace on the evening of November 9. The airline captain steers his craft into its designated landing approach. As he does so, he informs you with great relish of the immensity being displayed below: half the state of New Jersey can be seen to the left, Manhattan lies directly ahead, and the endless neighborhoods of Queens and Brooklyn are creeping into view on the right. The captain is pleased with the sights as much as any tourist, but he is of course professionally alert and regularly darts a glance down to his instrument panel. This time, when his eyes roll up again to reclaim his windshield view, all he sees is black. New York has disappeared. In one-half of a second, in a vertigo inducing reversal, the greatest metropolitan cluster of electrical lights in the known universe has been dimmed to oblivion. One pilot's exact thought was this: It looked like the end of the world.

With no landable airport in sight, he turns the plane into a new heading and seeks radio instructions on what to do next. From the standpoint of aeronautical safety, it didn't hurt that a full moon had just climbed into the crystalline early evening sky. Although nobody on the plane knew it at the time, in rough round numbers electricity had been snatched away from 30 million people in parts of nine U.S. states and a Canadian province over an area of some 80,000 square miles. In 12 minutes, 26,000 megawatts of power had disappeared from the grid when dozens of generators stopped singing.

UNIQUE NEW YORK

Defense. The authorities in Rome, New York, home of Griffiss Air Force Base, were thinking mainly of defense, national defense. Something big had happened, something electrical. Beyond that, it was a mystery. It wasn't merely a city emergency, but a major disruption affecting an important corner of the country. Therefore, higher priorities came into play. Is the president safe? Are the missiles at the ready if we need them? Did the dials show any hints of suspicious movements or hostile overflights anywhere remotely near the defense perimeter? What do the Joint Chiefs say? Had Defcon, the nation's Defense Condition, stayed at its normal level or been ratcheted up? Had there been any foul play? Were the Russians or Chinese mixed up in this?

The president of the United States, Lyndon Johnson, recovering from a gall bladder operation, is at the western White House in Texas. His press secretary, Bill Moyers, begins what will become a marathon news briefing, providing seven hours of information as it becomes available. The first order of business, after seeing to national security, is to investigate the cause of the electrical failure. The president, according to Moyers, has ordered the Federal Power Commission (FPC) to look hard into the matter, use all government services, including the FBI and Pentagon, and get back to him promptly. It is going to be a long night.

It is a mark of New York City's command on our attention that no description of the great blackout of 1965 could leave it out of account. In Buffalo, or Hartford, or Burlington, the local story might well take precedence, but the impact of the blackout on Broadway theaters or the 800,000 stuck in New York's subways would necessarily be part of the general impact of the blackout no matter where you lived, from Toledo to Tokyo. It's as if the seriousness of the event had been affirmed by the fact that it had befallen the Big Apple. If the lights can go out there, they can go out anywhere.

For many newspapers the main story wasn't the blackout—it was the blackout *in New York*. It's not hard to see why. The broadcast booths from which the national television networks would normally have reported the news of so large an event were themselves *right there* and victimized by the event. A third of the nation's overnight check clearing in the big downtown banks had not occurred. One-fifth of the nation's mail was left unsorted in New York's cavernous postal facilities. Because the power failure had struck New York, the blackout could unhesitatingly be deemed as officially "big."

Misery is companionable, and what better darksome companion to have than the number one city, brought down a notch or two. It was good to see that the center of the universe, sometimes reckoned by New Yorkers as being located at Times Square, could be traduced by something as ordinary as electricity. In the dark, the corner of 42nd Street and Broadway was no better than downtown New Haven or Schenectady. Yet even in the dark, there is something grand about New York. Here, all the good points and bad points of other cities come together and are amplified. And this proposition certainly extends to the deprivation and restoration of electrical service. Herewith a rundown of conditions in the moonlit but otherwise dark metropolis.

Except for the 30 million—the aggregate population of the dark realm, stretching from Buffalo to Boston—the blackout number that seems most to seize the imagination is 800,000, the headcount of those in transit through the New York City subway system at the fateful moment. Some were in stations at the time and could easily or brusquely jostle their way to the surface. Others were on trains in or near a station and could emerge relatively unscathed. The ones who weren't so lucky, the thousands in elevated sections or underworld tunnels, were sentenced to penitentiary periods stretching from minutes to hours.

For example, if at 5:20 p.m. that night you had boarded the IRT train (now the Number 7 Train) from Grand Central in Manhattan on its way toward the Vernon-Jackson Station in Queens, at 5:28 p.m., the moment of electricity's abdication, you would have found yourself in a dead halt in a tunnel beneath the East River. For two hours passengers sat in the feeble illumination of the car's emergency light before being led by police officers along the damp trackway the mile or so to the Queens-side station.

The thought of being in that train under that river amid that gloom for two hours is unpleasant enough, but now cast your imagination into another stalled train. Skim the slimy East River bottom a half mile to the north, and there inside a parallel tunnel beneath the silt and accumulated sunken, settled, castoff grocery carts, in its own tomb, an Astoria-bound train reposes. At the two-hour mark, when the passengers on the Vernon-Jackson train were being liberated, those on the Astoria train were just settling in for what would be a much longer stay.

THE ENTERTAINMENT NEWS

There's no business like show business, where the product is not any material thing but a dramatic situation, expensive sets, and lots of dancing legs, all burned into customers' minds with the help of 1,000-watt bulbs. With the theaters shut down and 30,000 high-priced tickets to refund, Broadway was particularly hard hit by the absence of electricity. The movie houses: total disaster. No electricity means no projection, and that means no movie. At Radio City, where they sometimes have a live organist accompanying the feature presentation, when the film being shown flicked off the screen, the music actually continued. Some viewers, with nothing to view and nowhere to go, spent the night. It was useless to ask, but millions did anyway. When was the power coming back on in New York?

At Carnegie Hall another trouper went on with the show. Vladimir Horowitz, rehearsing on the piano before an audience of students, was so practiced in his art that even after the stage went black, his knowledgeable fingers flawlessly finished Chopin's "Polonaise Fantasy." At the Metropolitan Opera, the show was to have been *Madama Butterfly*.

Instead of a stabbing followed by a blackout, there was only the blackout. A substitute performance went up a few days later.

New York was a hub for television broadcasting, an exercise that turns visual and audio information into waves and then beams them forth with a powerful transmitter from a large antenna, often on top of the Empire State Building. At the receiving end are 1965-era television sets—fat, power hungry, and often ensconced in a piece of mahogany furniture.

Not so with radio. With several of the sending antennas positioned conveniently across the Hudson River in still-lit-up New Jersey, many radio stations thrived. And with only audio signals to transmit, power requirements are more modest than for TV. Furthermore, radio receivers, could often fit snugly into auto dashboards. New handheld "transistorized" units were also popular. Because of this mobility, radio became *the* populist medium for passing the word, whether it be the status of Air Force bombers or the likelihood of the power coming back. *Variety* magazine summed it up:

> The day radio men dream about—when TV disappears and there is only radio—finally arrived without warning. Disk jockeys became news gatherers via telephonic quizzing of officials (governors on down), utility execs, and private citizens with graphic sidebar data to convey. Nearly all stations were yeomen in the breach, giving the AM medium one of its finer hours.[1]

Several hours along and power is returning in some places. Rochester has lights, and water filtration restarts. A preacher, representing the Negro community in Rochester, speaks out forcefully against the false accusation that there had been riots and looting in his community. Over in Massachusetts, where a real riot is making a mess of the state penitentiary, tear gas proves decisive. Although the guards in the death row part of the prison are nearly overcome by their own tear gas, backup officers from around the state have pretty much quelled the uprising. The lights are back on in Toronto and Vermont but not yet in New York City. Why is that?

An even more basic question, one that haunts engineers everywhere is: What caused the grid's downfall in the first place? Can it happen all over again an hour after we turn the lights back on? Are we raising a house of cards that will tumble with the merest nudge?

The Federal Power Commission has jurisdiction over the interstate movement of electricity and has already set up a mandatory meeting of utility officials for the next day. Representatives of the companies involved are to report on what they know, and a reconstruction of the disaster is to be undertaken. Placement of blame, the government assures nervous utilities, is not as important as preventing a recurrence. Changes in power policy will come later.

Between now and the morning, thousands of miles of transmission lines have to be inspected, circuit breakers checked, computer printouts scanned, and equipment cautiously turned back on. And before the cause can be located, the symptoms have to be found. Was it machine error or human error? Was malice involved? The idea of sabotage, although discounted early in the evening, never quite leaves the collective thinking. The rumors of utility men being murdered near Syracuse or of fireballs in the air just before the blackout are checked out and dismissed. Although somewhat reassuring, the fact that there seems to be nothing *physically* wrong with the grid is baffling. If there were no marks on the body, what was the cause of death?

The night wears on and most creatures great and small need a place to sleep. In New York's zoos, the nonhuman residents don't determine their own fate, so keepers have to make the appropriate arrangements. In a few cases, blankets are stuffed into the bars at the monkey house to preserve a bit of warmth. For some of the more difficult lodgers, such as the cobras, propane heaters are procured.

As for the Homo sapient residents of New York, many make it to their own homes, some after walking 10 miles or more—perhaps their longest hike since childhood camp outings. Others, in elevators and subway cars, haven't budged at all. For in-between cases—you're not exactly trapped, but without trains neither can you go the 30 miles to your home—you might make do with an office couch. At Macy's, which advertises itself as the world's largest store, thousands of customers are allowed to stay on for the night and are treated to an extemporaneous dinner in the employee cafeteria. Afterwards many settle into the home-furnishings department as if it were a huge hotel suite with accommodations for hundreds.[2]

It's 11:30 p.m. and Bonwit Teller, the jewelry store, hires two buses to take staff members out of town. So as not to get lost in the trek to the buses, they decide to hold hands. Maybe inspired by *Fiddler on the Roof*, playing over on Broadway (at least it had been the night before), the dealers in precious stones emerge hand in hand from their store singing and dancing.

PUTTING HUMPTY TOGETHER AGAIN

The *New Yorker* magazine, which prides itself on taking impressionistic notice of goings-on about town, certainly took notice of the great darkening. Its idiosyncratic coverage of November 9 concentrated on two qualities: the beauty and the fear. The fear, at least at first, was that the electrical disaster might be the start of nuclear Armageddon. This was, after all, the heart of the Cold War years. As the hours crept past peacefully, though, the fear dissipated. The beauty part was being able to see the silhouettes of the skyscrapers against the bright moonlight and the firefly blinking patterns coming from cigarette lighters and matches being struck by other lost souls visible in buildings across the street. This image nicely nails a central antithetical attribute of New York life, namely the illusion that your coinhabitants can appear simultaneously nearby and far off.

The *New Yorker* writer was, in the course of recounting his walk home in the dark, practically giddy in his sense of being cut adrift. Surely this was a nondisastrous disaster, a mere "festival of inconvenience," he suggested. Moreover, it was practically a pleasure to realize that things were getting on despite the fact that no one knew what was going on, not the mayor, not Con Ed, not the president. Earth, or at least our technologically built-up part of the planet, was being taught a lesson in humility, as if by some omniscient other-worldly intelligence.[3]

Actually, the hard-working crews of Consolidated Edison *did* know what was going on, or shortly would know. Yes, there were complaints. Rochester had largely been back up since 7:30 p.m. and Toronto an hour later. Why not New York? Several good reasons. First, when the world's most complex urban power network comes unhinged, you don't just turn it back on with the throw of a switch. The city's power stations work not by simply converting falling river water into electricity. No, the New York plants, all fire and brimstone, use steam at a thousand degrees. Steam is a dangerous fluid requiring careful handling. It's a high-pressure, high-temperature, high-voltage, heavy-metal environment, and all the parts would have to be inspected from the inside out. Before you could put Humpty Dumpty back together again, you might even have to break him into smaller bits. You have to make sure that everything is truly *off* before you start turning anything back *on*. Not only does New York have to be isolated from the rest of the Eastern Interconnection, but the boroughs and all the 42 separate sector networks have to be severed from each other in order, later, to form a more perfect union.

Before taking heat again, furnace valves must be oiled. The roundness of all those spinning turbines has to be verified before they can be set rotating again. The monstrous machines swirl so much weight around so quickly that even the tiniest imbalance would cause the turbines to tear themselves apart. You know the racket an imbalanced washing machine can make? Amplify that by a million.

Mostly this scurrying activity by utility workers is invisible to city dwellers with problems of their own. They're stuck in an elevator, or trying to hitch a ride home, or attempting to hire at extortionate rates a taxicab ride out of midtown. Schrafft's loses $200,000 worth of ice cream. Meanwhile, out in the bay, light beams so gallantly streaming continue to emanate from the Statue of Liberty, which gave proof through the night that it was hooked up to the Jersey side of the bay, not the New York side.

The electric grid might not be working, but the telephone grid is pulsing, owing to backup generators. Therefore, a million calls go through, a record number. Where are you? I'm stuck in the office. When will you be home? No way of telling. When are they turning the power back on? No one knows.

Establishing the story of what happened, not just resuscitating the machines, is an important duty. Journalism proceeds in tandem with electrical engineering. Some of the blackout incidentals related here come from government or utility reports. Many more come from daily newspapers, those great, buoyant diaries of town life. Each edition of a newspaper is a time capsule of that day's activity. Stories, headlines, pictures, advertisements—all are valuable in vividly revealing the existential inventory of a city's physical reality, the particulars of domestic, artistic, political, and economic life. It is all food for thought and becomes one of the most tangible ways, decades or centuries later, of telling what it was like then. Even if the participants of those events could somehow be kept alive and made available for questioning, the accounts would still be untrustworthy owing to the inevitable drift of fallible memory. Therefore, the ladies and gentlemen of the press necessarily become the abstract and brief chronicles of the time..

In New York City all the morning newspapers but one decided to give up trying to get out an edition. The one holdout, the so-called paper of record, the publication where the staff take the news, and themselves, very seriously, declared that it would be full speed ahead. Setting aside the originally planned 96-page spread, the managers at Times Square decided to trim back to a slim but practical eight-page version featuring, naturally, the story that was going on all around them. The other stories of the day were going to be there too but would take backseat to the unprecedented technological mishap.

These tribunes of the people, the writers of the *New York Times,* worked that night by candle light. By 1965 all the newsroom gaslights were gone, so it would have to be candles or nothing. But where does one procure candles enough for a whole newspaper staff? A few are garnered from the Astor Hotel and some from another great New York institution, the Catholic Church—in this case, Holy Cross and St. Malachy's.[4] There is still, however, a daunting lack of mechanical power. You can write by candlelight, but you can't run a printing press powered by wax. To publish a paper by morning, something else would have to come along, maybe a miracle.

You can't believe everything you read in the papers. For example, correspondents from the Soviet Union couldn't get the story straight. *Tass* reported that the streets of New York were relatively quiet and orderly, while *Izvestia* reported hysteria; politicians and citizens alike were said to be in a state of panic. Reports from British papers quoted utility executives in the United Kingdom that such a blackout could never happen on their grid. To readers in France, whose nationally owned grid was said to be flawless, it appeared that the Canadian-American failure sprang from reluctance by stingy privately owned utilities to invest in the right equipment. In Germany a more far-seeing engineer suggested that a completely reliable power system was impossible to achieve and that local officials should prepare for the day when it would happen in Germany. Photographs of New Yorkers camping in Grand Central Station were fascinating to Londoners, who only 20 years earlier had been forced to rough it in that way during the Blitz.[5]

On the evening of November 9, electrical engineers in the United States were not worrying about the grids in Germany or France, but they were thinking of their counterparts in Canada. Utility officials on either side of the U.S.–Canada line had the early impression that the blackout had begun on the other side of that long border, and the pursuit of the truth would have to be handled gingerly. Representatives from all the affected utilities would converge in Washington, D.C., with logbooks and possible explanations at the ready. According to the chairman of the FPC, Joseph Swidler, he received mostly cordial cooperation that night, except for Consolidated Edison, which did not at first want to divert its logbooks from New York to Washington. When Mr. Swidler mentioned that the president of the United States had assured him of unlimited access to agencies of the federal government, including the FBI, Con Ed quickly agreed to deliver its books.[6]

As the evening hours wore on, there was, among all the contributories to the Eastern Interconnection energy pool, still scant evidence of any physical damage. For an event that wrenched 30 million people back into the Dark Ages, you'd think there would be some clues of violence somewhere. Actually there had been some damage done, serious damage, but this had been the result, not the cause, of the blackout. The news from Queens was not good. One of the big units at the Astoria station was damaged, another at the East River station. And when the name of Ravenswood Unit Number 3 was read aloud, the worrisome engineers at Con Ed groaned. Big Allis, the company's most valuable workhorse, had been wounded in action. With these three generators, all of them lying lifeless along the East River shoreline like beached whales, Con Ed had lost a fifth of its capacity to generate electricity, an amount exceeding the power-making ability of most cities.

With a mighty heart, Big Allis had valiantly tried to sustain what was left of the Eastern Interconnection in its dire moment. It had raised more than 100 megawatts in the handful of minutes before the final avalanche. At the core of the leviathan, the 540-ton, 145-foot-long rotating shaft had been crippled when its bearings burned out. The lid on the turbine itself weighed 150 tons. To extract the tension bolts holding the whole thing together, gas torches had to be used to warm the metal. The doctor's prognosis? The patient would be disabled for at least a month.[7]

But what had been the cause of it all? The generalized forensic inquiry would proceed down into the grid's lower depths if necessary. Since no visible damage had been found, investigators would look at the second-by-second printouts of the devices that monitor voltage levels and power flow in all the important passageways of the network. The sleuths were confident they would pluck out the heart of the mystery. Like deducing an arrhythmia of the heart by deciphering excursions on an electrocardiogram, so the arrhythmias of the grid could be gleaned from the seismic waveforms representing the critical high-voltage levels at or around 5:16 p.m. on the night in question. This particularized remembrance of electricity past, everyone believed, would reveal possible sequences of causality.

The disruption certainly had manifested itself at an early moment on the transmission line between Rochester and Syracuse, but this had not been the epicenter of trouble. The true cause, the experts later learned, lay farther west, right at Niagara, where the two friendly countries come together with power plants on either side of the river in an international spaghetti entanglement of high-voltage lines. To triangulate the cause of the crash and propose a remedy would, in the coming days, require persistent investigative work combined with sensitive diplomacy.

CRITICAL BEHAVIOR

Achieving a grid that never crashes is like trying to reach the speed of light or a temperature of absolute zero. It can't be done. Planets, baseballs, and human bodies must all obey the laws of nature, and the electrical grid is no exception. And one of nature's central themes is decay and death, not just for living organisms but for all composite things. Even mountains and stars fall apart.

One way to study this falling-apart process, particularly in complicated systems like the electrical grid, is to look at what happens to sandpiles. Yes, something as seemingly simple as a mound of sand is useful in modeling behavior in an electrical network draped across a thousand-mile-wide landscape. Experiments with sand, and other artificial networks such as arrays of tiny blocks fastened together with elastic springs, are an integral part of a relatively young discipline called complexity science, which tries to understand systems containing many rapidly interacting parts. Knowing the detailed behavior of individual parts—circuit breakers, say—isn't enough for you to understand the whole system. Behavior of the whole *emerges* from complicated reactions among the components. In other words, the whole is greater than the sum of its parts. This emergent behavior of the whole can be unpredictable at times but isn't necessarily unexplainable.

To see how complexity works, take a closer look at a sandpile. If you add grains to a cone-shaped pile one at a time, mostly nothing happens. Sometimes an extra grain—plop—will cause a small avalanche here or there for a second or two. Then everything gets quiet again. Another grain is added, and not much happens. Conditions are gradually changing though, even if you can't see them. Inside the pile the grains communicate with each other through subtle nudges and scrapes. With every new grain added, the internal ordering of the pile gets readjusted. Stability is never more than provisional as the conical shape gets more precariously steep.

Now and then—plop—a new grain will cause a more substantial portion of the pile to give way, after which a new stability is established. Then the process begins again. More grains are added—plop, plop—causing more internal adjustments. Lots of small avalanches happen and some medium avalanches. Fragile stability is always heading for a larger instability as the pile gets steeper and as the angle of repose becomes more acute. The system, the sandpile, is moving toward still another avalanche event. The placement of one tiny grain can trigger the placement of all the others. This system of thousands or millions of parts can be placid on the outside while moving toward catastrophe on the inside. Finally, the big one arrives. Plop: A fatal grain is added, avalanches of all sizes occur simultaneously, all sense of scale is lost, and the whole pile comes crashing down.[8]

Here comes the big generalization. The electrical grid and its thousands of rapidly interacting components can resemble the behavior of a sandpile.[9] It's as if the grid, and maybe the rest of interconnected modern technology too, were built on the side of a steep slope. The grid is subject to a steady plop-plop-plop of irritations—frayed insulation, lightning strikes, misbehaving generators, a surge in air conditioning—which can trigger avalanches in the form of blackouts, some small, some large. The precipitating event can even come from space; eruptions on the sun occasionally spew potent streams of charged particles that, when they push into our atmosphere, can induce voltage fluctuations that disable satellites and even disrupt electrical grids on the ground.[10]

Studies have shown, furthermore, that the criticality behavior of the grid is shared by a number of other large complex systems prone to disasters such as earthquakes, river floods, hurricanes, and forest fires. If you make a graph showing how often disasters of a certain size occur, taking into account the differing ways you measure disasters (megawatts lost for blackouts, destructiveness for hurricanes, acres burned for forest fires, and so on), the curves for these complex systems are nearly identical.[11] The grid is obviously an artificially engineered thing, while these other systems are natural. Nevertheless, what they have in common is a quick interactivity of component parts, whether it's warm air masses joining forces in the Caribbean (in the case of hurricanes) or transmission lines floundering in Ontario.

How did the grid's complexity and criticality turn into the specific electrical disaster that happened on November 9? After all, the power humming through the wires of the Eastern Interconnection didn't just disappear down a rabbit hole on that day. The customers wanted to know, the engineers needed to know, the utility stockholders insisted on knowing what went wrong. No one would relish having to apologize to the 30 million. Investigators uncovered many imperfections: faulty system architecture, poor procedure, errors in judgment, insufficient readiness. But was there a single locatable origin to the disaster?

THE PARTICULAR FAULT

Yes, there had been a fateful grain of sand. Its arrival on that gigantic pyramid of electrified complexity we call the Eastern Interconnection had nudged things beyond the tipping point. The dead center of the fiasco lay placidly inside a vault at the Sir Adam Beck hydroelectric power plant in the town of Queenstown, Ontario, not far from the Niagara River. Operated by the Ontario Hydro Power Company, the Beck facility supplied its own customers to the north and west on the Canadian side of the border and also shipped huge blocks of power south and east to the U.S. side.

Rerun the phenomenon. For the moment, picture a set of five metal cables extending from the Beck plant all the way to Toronto, the largest city in Canada and a prime load center for electricity. A cable carrying electricity is a simple thing. It has no moving parts and does one thing—carry energy. To ensure that the cable itself and the attached assets such as generators and transformers are protected from burning out, circuit breakers are ready to interrupt the flow in a fraction of a second by retracting a large metal contact.

What happens when a line is opened? At first the electricity isn't fully aware of the interruption. Like a stream of cars driving into the river after a span of a bridge has collapsed, the electrical current is inclined to continue flowing, at least for another instant. It does this in the form of an arc, a miniature lightning bolt leaping across the now-widening gap. A high-voltage circuit breaker provides muscle, sinew, and pivoting joints. What it lacks is a brain. Brainpower, or at least the sensory capability to discern a change in the local environment and then do something about it, is supplied by a device called a relay. It tells the breaker when to flash into action. Like expressway on-ramp traffic lights that sense when it is safe for a car to enter (green light) or when it is too crowded (red light), an electrical relay continuously monitors traffic conditions and allows current to continue or to be cut off.

If, as in the old days, we believed it was irony-loving gods who set disasters in motion, you could not ask for a more appropriate chain of disproportionate causality. The misbehavior of something small, no bigger than a breadbox and containing no explosive or deadly toxin, could disrupt the lives of 30 million people. The twig that broke was a safety device, one of those relays designed to limit the flow of current that inadvertently unleashed ruinous amounts of current. The nearest parts of the grid would generally be affected the least, while the hardest hit would be the parts farthest from the source.

Here specifically was the problem. Those five cables pointed at Toronto were doing fine and carrying current well within their ratings, but the relays standing guard had been set too low to handle the larger amounts of power being imported from the United States. Without anyone knowing it, the Ontario grid, and along with it the whole Eastern Interconnection, had been moving ominously up a steepening slope of instability.

The different parts of this collective sandpile had been organizing and reorganizing for months, maybe years, toward a condition of greater vulnerability. The parts, being made of mute metal, do not communicate with each other in the meaningful way that humans do, but they *do* communicate. They are in physical contact with each other. We don't normally think of the numerous grids making up the Eastern Interconnection as being a single machine; it took the events of November 9 to demonstrate this singleness in a spectacular way. If the avalanche didn't happen on this day or the next, it would happen later. The day finally arrived. The decisive grain of sand fell onto the pile. The toboggan ride was about to begin. This then is the tale of how electrified civilization came unzipped on November 9.

At 5:16 p.m. Toronto was in the middle of its rush hour, and a slight upward tick in the current drawn from the Beck plant was enough to actuate the relay, which quite properly (it was only obeying orders) signaled a circuit breaker to open, taking the line out of service. Denied the use of that route, the energy adroitly detoured to the four companion lines. Not surprisingly, the relays doing sentinel duty on those lines also hastily determined that too much current was flowing. Breakers on *those* lines were now activated and suddenly, in a matter of a few seconds, all of Beck's Toronto-bound energy was stymied.

This frustrated power now turned about and headed for New York. It tried to get to Toronto the only way it could—by going the long way counterclockwise around Lake Ontario. So a huge wave of electricity came surging down into New York state in addition to the current that was flowing there already.[12] This tsunami of energy threw the generators to the south into asynchronous confusion. They no longer spoke the same language, and when that happens emergency procedures kick in. Relays everywhere went crazy, lines opened, generators tripped off, cities went dark.

This was the too-much-energy part of the blackout. Then, with most of the power from the Niagara end of the Eastern Interconnection blocked off because of open circuit breakers, the not-enough part of the blackout began in all the other places. With the general network now undergoing fibrillation, the avalanche dynamics dictated that localized portions of the grid should scramble for extra power wherever they could find it. In the minutes after the Beck failure this meant, in practice, the power plants of the Consolidated Edison system.

The most put-upon machine that day was Big Allis. You have already seen what happened next and last—how a cascade of shutdowns elsewhere put more strain on this overworked machine, how it lost all its own energy, how its internal oil pumps neglected to work, how its bearings ran dry, how its central shaft continued to rotate anyway at 3,600 rpm, and how, deprived of its lubrication,
the twirling slowly, slowly came to
a quite unsatisfactory
and grinding
halt.

MEGAMACHINE

Thirty million people without power! How had it come to this? Lewis Mumford was not surprised, but he was disappointed. He had hoped the age of electricity would help liberate people from drudgery. Instead they seemed to be as mired as ever. Moreover, our general enslavement to machines was not diminishing but accelerating. We were now seeing the "automation of automation," he said.[13] A mechanical relay at a Niagara power plant, deciding on its own, opened a circuit breaker when it shouldn't have, while a computer in Manhattan *should have* opened a circuit breaker but didn't. The result: a hundred cities fell apart for a day. Let's leave the lights off in Manhattan just a few minutes longer while we ponder this state of affairs.

Through his books and his reviews in the *New Yorker*, Mumford kept updating his opinion of technology and its impact on city life. When he began to write in the 1920s, the grid was then just coming into bloom. By the 1960s, Mumford felt compelled to take more factors into account. Previously he had extended our sense of the "machine age" backward from the electric-automobile-radio age to the steam-coal-iron age and from there back to the first mechanical clocks of the 13th century. Now, in order to make his technological theory of everything more inclusive, he pushed the origins of machine culture further back to the time of the Pharaohs.

Here was his line of reasoning. The arduous construction of the Egyptian pyramids had not used geared metal machines. But in the exactitude of the surveying, in the precision of the stonecutting, in the ingenious transport and placement of the blocks, and in the masterful management of so colossal a project, it would seem a machine *had* been involved. This machine, Mumford argued, was made of human parts. The pistons of the machine consisted of human arms and legs. The gears of the machine consisted of human joints. The energy needed to sustain the machine consisted of the rations meted out to workers at feeding time.

Furthermore, the totality of this 100,000-person machine included not just the laborers but the surveyors and engineers, an army of priests, scribes, astronomers, and soldiers for enforcing order. Mumford referred to the whole mechanism—people and tools—as a "megamachine." Wanting to draw dramatic attention to what he saw as the regimentation of our own day, he asserted that we too are all parts in a modern megamachine.[14]

Mumford was not suggesting that New Yorkers had anything like the living conditions or political status of Egyptian workers in ancient times. But he did want to highlight the similarities between the pyramid project on the Nile and the fully staffed corporate skyscraper on the Hudson. One big difference between the two was that the pyramid was dedicated to a king, while the doings inside the skyscraper were dedicated to the corporate megamachine itself.

Just a minute. Are we ants? Are we merely unwitting and unthinking parts of a computerized mechanism running on electricity? Well, if we stand atop one of those skyscrapers on the evening of November 9 what we might see, looking down from above, would resemble something like an anthill that had just been given a swift kick. We would see long columns of tiny creatures, issuing from the now functionless buildings, and then scurrying along the ribbons of the avenues, heading on foot toward some distant borough (or burrow).

This analogy is of course unfair. City workers are not ants or cogs operating according to a unified plan. Each New Yorker is unique. Even a confirmed grid builder like Samuel Insull had counted on the multipurposed nature of his customers (". . . So varied are the ideas of human beings, and they so seldom do the same thing at exactly the same moment . . .") to smooth out the load curve for his utility. Admittedly, most residents of New York, or any other city of the world, are unable to see the Milky Way from their back windows, and many parents, if asked, would admit that they ought to read more often to their children, and very few nowadays are able to entertain themselves or their friends by playing a piano in their parlors. But it's hard to know what Mumford would like us to do about all of this. New York City is not Walden Pond.

Mumford's megamachine theory only crudely models the social dynamics of actual cities, but it does, I believe, nicely anticipate the complexity theories developed decades later. In other words, I see some resemblances between, on the one hand, his descriptions of entrenched regimentation, mass production of goods and services, and the general (if not fully informed) consent and cooperation of customers with manufacturers of goods in creating popular economic and cultural life and, on the other hand, the hair-trigger relations existing among the rapidly interacting parts of a sandpile-like city complexity producing perpetual realignments and occasional chain reactions. Is not our megamachine technology characterized by faster connectivity (nanoseconds now, picoseconds next year) and greater informational loading (gigabytes now, terabytes next year)? What is an urban population but a mound of individual (although nearly identical) units rubbing together? What is fashion (in clothing, automobiles, television programming, engineering) if not a series of small-scale avalanches in taste triggered higher up the crumbling slope?

Avalanches are inevitable in a system of growing complexity. To use again the sandpile analogy, society can erect retaining walls to hold back, for a while, the sandslides. But as more grains are added (plop, plop), as steepness grows toward a critical point, the onset of larger avalanches becomes nearer. Artificial levees will work only for so long. The sand will find a way, over the top of the levee or through a crack, to flow downhill. Should we build levees? Yes. Will they reduce the threat of large blackouts to zero? No. Because our society insists on building its technology on an ever-steepening hillside, we should expect avalanches.

FIT TO PRINT

November 9, 1965. The hurricane on the internal energy sea had come and gone. In many cities the storm's visitation was short lived, in some cases for only an hour or two. In a few places the power had not failed at all since someone had adroitly intervened at a decisive moment or because an automatic switch had done the job. Hartford was one of these fortunate islands of light, owing to the splendid stand-in performance by an old power plant, affectionately known as Old Gal, normally kept in service now only to help meet the evening peak power demand. Because it did not hook into the regional grid, as the other generators did, but only into Hartford itself, Old Gal remained up and running. She had kept her head of steam even as others were losing theirs.

Things weren't so providential in Rhode Island. There, the local utility had no provision for auxiliary power, the thinking being that they would never run completely out of electricity. The Narragansett Electric Company therefore had to wait for help from outside, which arrived from a generator in Millbury, Massachusetts. With this start-up electricity in hand, Providence began to restore energy in a process that repeated itself all over the Eastern Interconnection. Cautiously turn up the generation, deliver power to one sector, monitor the frequency and voltage, crank up the boilers some more, increase still further the current going into the large copper conductors (the bus bars) leading out of the plant, light up another neighborhood, and so on. If the power wasn't quite enough for the load, some feeders had to be turned off, at least temporarily, then on again when the electricity *was* available. The golden rule always to be observed: Supply and demand must be in balance. In Providence, though, they were going for broke. Five main lines would be activated all at once. Dispatchers were waiting at their switches and then came the count: One, two, three . . . and voila! The circuits were closed, power ran its course, and the city had light again.

How about New York City? Why didn't it have power? All through the night exasperated New Yorkers were asking this question. Because of an oddity in the way the power failure had arrived, Staten Island and a sliver of Brooklyn had been left active. This was to be the lifeboat of the New York system. From here the other parts of the city would be resuscitated, starting with the darkened parts of Brooklyn. The U.S. Navy had offered the electrical resources of one of its ships lying in the East River, but the power wasn't needed.

And so, as it had in other places for hundreds of miles around, the procedure was repeated. Reignite the boilers, build pressure, open valves, and let steam come full tilt at the turbines. With electricity now makeable, neighborhoods would sequentially get back their due ration of energy. The reenergizing went according to plan with all deliberate speed. But to those on the outside, the restoration seemed delinquent. At 1 a.m. still only 25 percent of the city had power. Why did Albany or Toronto or Syracuse have energy but not the borough of Manhattan or the Bronx? Because there was much *more system* to be activated—hundreds of miles of high-voltage cables filled with a Niagara's worth of energy—the largest underground network of wire in the world.[15]

All the news in New York that was fit to print at that hour was about to be printed. The fullest account of the blackout, at least the civilian version of the affair and not the official federal report that would be assembled by engineers, bureaucrats, and utility officials a month later, was coming together at the offices of the *New York Times*, where three dozen newsmen were straining toward their self-appointed, publish-at-all-costs goal. And it would come at great cost since in the pared-down edition coming into being there would be little room for ads. The paper would go out practically gratis, all for the greater good.

As the evening hours wore on, it seemed more and more likely that the *Times*'s own press would be unavailable. The facilities of the *Newark Evening News* were therefore secured for two obvious reasons: First, they were located in New Jersey, which was blessed with power. Second, it was an evening paper and wouldn't be needing its press for the next few hours. Text for tomorrow's edition of the *Times*, some of it still unedited, was shuttled beneath the Hudson River through the Lincoln Tunnel. In all, some 21,000 words were shipped west.

Newark not only gave the *Times* a press to play with but also a set of complete Wall Street stock closings, so at the last minute the *Times* edition was bumped up to 10 pages. After all, business is business, and even in the middle of the greatest blackout ever, it would be nice to know what your holdings were doing on November 9.

Finally, Turn On. At the Waterside plant, the oldest continuously serving power station in New York and the energy provider for the heart of Manhattan, things were moving to a climax. The plant's apparatus was inspected, steamed up, and ready to go. Now it was time to move power back to where it was needed, in the homes of New Yorkers. It was time to pick up load.

To do this, the old-fashioned method was adopted, just the way it had been done in Providence a few hours before. To shoot energy back into circuits, the engineers had to take the system into their own hands, literally. It had all come down to this single moment of truth. Here, according to an engineer on duty at the time, was the scene in the wee hours of the morning:

> After the station bus was energized, 12 men were stationed at the high board and were told to take a switch in each hand. With controls for each of the 24 feeders thus in hand they were instructed to operate the switches instantly at the count of three . . . 24 switches slammed into closed position, instantly transforming the Grand Central network from a sea of blackness to a sea of light.[16]

Everyone who didn't have a hand on one of the switches watched through the windows. When the energy went through, the results came back at light-speed. The city was relit! Grand Central was grand again. Relieved, the men let out a cheer of delight. The city was almost back from the dead.

What about those poor souls on the Astoria Train trapped beneath the river? Some of them had walked out on their own steam, traipsing up the track toward the Queens side. Others, afraid to make the long tunnel walk or worried about emerging into a neighborhood they didn't understand in the middle of the night without benefit of street lights, had decided to sit tight. How long did they sit? When the power finally came on for that train, when things finally moved, the time was 7:15 a.m., almost 14 hours after the train had launched itself into that tunnel the night before.[17]

Not all the blackout narratives had happy endings. Vange Burnett, in town from Florida for a trade show got lost in the corridors of the Windsor Hotel. He did not show up the next day, nor the one after that. Could he have just left for home? No, because the family hadn't heard anything either. Sad to say, his dead body was discovered six days later in the pit of his hotel's elevator shaft. The hand of the dead man still clutched a candle.[18] Perhaps the candle had gone out and the man had taken a fatal fall in the dark.

There are a million stories in the naked city. If you had wanted to delve just beneath the surface reality for many of the named figures in this book, such as Tesla or Westinghouse, you would find whole books. For Vange Burnett there are no books, but there are a few Internet websites where, for example, you can learn that he served as a lieutenant commander in the U.S. Navy during World War II. You can see Vange's high school discus and shotput records enshrined for the year 1931. On other pages you can view his picture taken at a Northwestern University football commemorative, or his place in a Burnett family tree that proceeds lineally downward to grandchildren now alive and upward to forebears preceding him by a hundred years or more.

The great blackout of 1965 was over. When the engineers had finished their valiant job of correcting electrical faults, the editorialists began their job of finding fault. *Life* magazine declared that the crisis had had its good points since "it deflates human smugness about our miraculous technology." The dark skyscrapers, glimpsed against the fading sunset glow over New Jersey were, *Life* said, "as lifeless as the hulks of Angkor Wat."[19]

The *New York Times* made it to the streets even as the last shreds of tardy voltage were toddling into place in the grid. The 400,000 copies quickly sold out at corner newsstands the morning of November 10 and became keepsake possessions to preserve in lower drawers of attic wardrobes. Not only did the *Times* detail the events that had unfolded just a few hours before. There were also, as a bonus, those closing stock quotations for November 9.

How about quotations for November 10? What did investors make of the largest ever single-day technological outage? Well, the New York Stock Exchange did open (one hour late), and traders quickly told the manufacturers of electricity what they thought of the night before. Consolidated Edison was down 3/8, Niagara Mohawk 3/4, and Pennsylvania Power & Light 1/8. Meanwhile, the makers of heavy duty power equipment did better: General Electric was up 15/8 and Westinghouse 5/8. A circuit breaker company's stock closed even higher.

7

OVERHAULING THE GRID

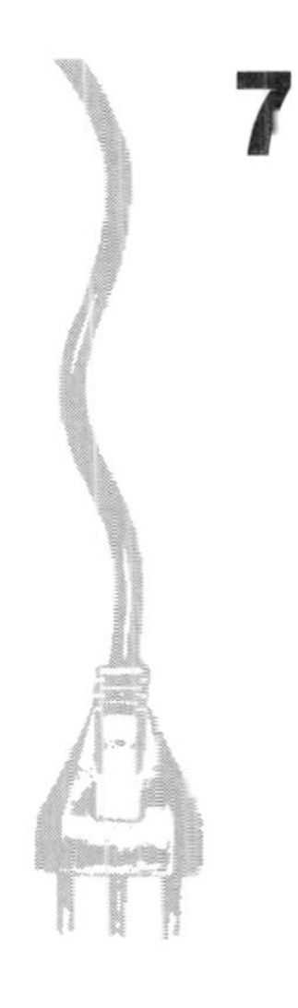

> Well, it's no trick to make a lot of money . . . if all you want to do is make a lot of money.
>
> —*Citizen Kane*

Utility stock wasn't the only thing down the morning after November 9. So was confidence in the idea of shared electricity. If a problem in Rochester or Toronto could ripple all the way down to New York City, why don't they just cut the connecting wires? But that's not what happened. Not only did the connecting wires stay where they were, but new ones were added. The energy sea would in coming years become more like an energy ocean.

Nobody would have said, the day after November 9, that the course of the grid business was about to swerve. Grid history, like all forms of history in the making, is hard to assess while you're living through it. It might be easy to see changes being made—the construction of new power plants, say—but difficult to fathom the meaning of the changes until later.

Still more difficult to achieve is an appreciation for the irony of the situation. The Greek playwrights made irony the centerpiece of their depictions of the mystery of life. The fundamental human predicament they wanted to illustrate was the following: People intend to do one thing, but often something very different happens. They pursue a course of action and then are swamped by unintended consequences. Surface appearance turns out to be very different from the underlying reality.

Irony is a part of virtually every human endeavor, especially the management of complex technology. This chapter in the history of the electrical grid, devoted to the aftermath of the 1965 blackout, is rich in irony. It is a tale of missed opportunities and occasions when a problem's symptoms were misread by those in authority, who then took a course of action that only compounded the problem. One of the great lessons of life, and of Greek drama, is the realization that no matter what you've been told or come to believe (and here I quote from that great American drama, George Gershwin's *Porgy and Bess*), "It ain't necessarily so."

NOT ENOUGH GRID

Ever since the time of the grid's founding fathers, power plants had been getting bigger. This was economy of scale: The larger the machine, the more output you got per unit of input. The cost of a kilowatt-hour kept tumbling. For the better part of the century, economic growth in America, in Germany, in Japan, in Argentina, benefited from electrical innovations. Graphs showed that the more electricity a place had the greater its prosperity. Greater efficiency and lower prices—they would just keep going forever, wouldn't they?

All the studies had shown that bigger was better. Bigger boiler, bigger pipes, hotter steam, higher pressure. All the engineering experience had pointed to what seemed like the inevitability of the Big Allis approach. If your utility could afford the financing or had a big enough revenue stream or workforce or experience to grapple with a machine of such monstrous proportion, you'd want a 1,000-megawatt source of electricity too. This would seem logical, but as we will see, it ain't necessarily so.

The immediate consequence of November 9 was an effort to make sure such a disaster didn't happen again. Improvements in sensors and transmission lines were mandated. An organization, the North American Electric Reliability Council (NERC), was created to foster growing regional ties among utilities and to ensure that prudent safety measures would be observed.

Prudence might have been the watchword, but something more pressing, an ineluctable factuality, was lurking inside the pipes and valves of the big power plants. This intestinal disorder, ignored or not noticed for several years, was soon to lead to major complications. The problem was this: Bigger was no longer better. There weren't yet enough data points to clinch the case, but it would become evident that a bigger boiler wasn't necessarily a better boiler. Efficiency had reached a plateau. If you factor in the maintenance downtime for the biggest generators, efficiency was actually on the decline.

The soldiers of the grid at the platoon level—the electricians, plumbers, pipe fitters, welders—could see firsthand that the big machines were troublesome. For example, at the extreme pressures and temperatures needed to operate the battleship-class machines, corrosion seemed to occur sooner. Components might do well in lab tests, but running continuously for months at unprecedented high temperatures took its toll. A failed part, even a small one, could mean that the whole 1,000 megawatts would have to be taken out of play. A bridge with four lanes can carry twice as many autos as a bridge with two lanes, but a single overturned truck can still shut down either bridge.

So it was with Big Allis in the late 1960s. The wisdom of bigger is better became questionable. Especially in those summertime heat waves, when the living is anything but easy, the flagship of Con Ed's fleet was chronically ill. And with so many of the company's eggs nestled in one large basket, everyone started to sweat on those days whenever Big Allis sneezed. After years of successful ads extolling the benefits of electric air conditioners, dishwashers ("Roll in a dishwasher . . . roll out work"), and clothes dryers ("You can't always be sure of sunshine, but you can always count on a clothes dryer."), the utilities' efforts to fulfill electrical obligations with existing generating capacity came to be more fraught with danger. Isn't it ironic? Having created aggressive consumers eager to use more electricity, the utilities were now having trouble keeping up with demand.

On the seesaw of power delivery, what you have on the load side—all the customers and their plugged-in appliances—must be exactly balanced every second by the generation side—all the current shooting out from all the plants. What happens, though, if a power plant is disabled, and the new plants are not yet installed, and the high today will be 95 degrees, and the company has used up all its extra reserves, but the customers keep asking for more?

At a moment of stress, one thing the electric company can reluctantly do is lower voltage. This is referred to as a brownout. Not as drastic as a blackout, a brownout is an anemic condition in which the regular 120 volts might be allowed to deteriorate by a few volts. Appliances don't like this lean diet. Motors will struggle, as if having an asthma attack. Incandescent bulbs, designed for 120 volts, will emit only half their normal light at 100 volts. Air conditioners can overheat, and subways go slow.

Let Consolidated Edison serve as the example of late 1960s grid operation. On some hot days, when their own generators weren't quite up to the task, Con Ed dispatchers went searching for extra power wherever they could, to Long Island, to Canada, even to Tennessee and the TVA.[1] The company was hardly standing still. In fact, it planned an ambitious program of expansion. For example, it sought permission to build a nuclear reactor at the Ravenswood plant, home to Big Allis, but was turned down after much rancorous debate.

An even more contentious battle swirled around Con Ed's proposal to build a pumped storage facility at Storm King Mountain on the Hudson River. Imagine moving a small piece of Niagara Falls a few hundred miles closer to New York City. It worked like this: At night, when demand was low, electricity would be used to pump water up the slope of a mountain into an 8-billion-gallon reservoir. The next day, at times of peak demand, the water would be allowed to pour back down off the mountain, regenerating electricity as it fell. True, you would recoup only about two-thirds of the power you used to pump the water uphill in the first place, but it was worth the effort since the rush-hour power would be *premium electricity*. It would be a supply of energy when you needed it most. If Storm King had existed earlier, the company argued, the 1965 disaster might have been averted. The brownouts that came later might not have been necessary.

Pretty much used to building plants whenever and wherever it thought fit, Con Ed was taken aback when a number of conservation organizations, such as the Audubon Society and the Sierra Club, joined local critics in combating the plan in court on the grounds that it was bad for the scenery, bad for fish and fowl, and bad for property values. The company was frustrated. Not only did the delays hamper their long-term planning, but it brought Con Ed bad publicity in the newspapers. Furthermore, the case galvanized many who had previously not been particularly vocal when it came to conservation matters.

Blocking Storm King was seen by some as a pivotal step in the evolution of the environmental movement.[2] The critics who had found their voice and the appropriate methods for standing up to the giant corporation would hereafter be a regular part of the culture of electricity, whether it concerned the siting of transmission lines, the release of sulfur into the sky from a smokestack, the killing of fish by heated water returning to a river from a power plant, or the falling of coal particles into lungs and onto property. A ton or more of dust sprinkled New York City's Central Park every day.[3] Things that used to be normal operating procedure were now being questioned.

The critics of the critics liked to ask how all the expected improvements, the nonpolluting generators and repairs made without digging up streets, would be made without also raising rates. Where, the utilities wanted to know, did the citizens and elected officials of the city want the needed generators to be built? Answers were often lacking.

Con Ed became the company that people loved to hate.[4] It was seen as an arrogant, indifferent, uncaring institution. What used to be held up as virtues, such as the habit of family members—fathers, cousins, nieces—working side by side or managers staying loyally at the job for 20 or 30 years, were now viewed as points of detraction. The company, some said, was out of touch with customers, and their equipment was antiquated. The *Wall Street Journal*, interviewing a consulting firm that had looked over the utility's management style, reported that "a survey of the 240 top executive jobs at Con Ed found that the duties of 71 executives consisted solely of supervising one other executive each."[5] Late-night television talk shows offered plenty of Con Ed jokes. The mayor, John Lindsay, adroitly refrained from working too closely with the main energy provider for his city. To an ambitious politician, having Con Ed as an opponent was much more advantageous than having the company as a partner.[6]

Con Ed vowed to make customer service its highest priority. It reduced the payroll, changed its motto from "Dig We Must" to "Clean Energy," and changed the color of its trucks from orange to the blue-white-gray scheme used to this day. The company responded faster to reports from homeowners of power troubles. The promotion of energy-gulping appliances was ended, and a "Save a Watt" program urging customers to use 10 percent less electricity in the summer was initiated. That's right—the power company was actually encouraging people to buy less of its product. Helpful hints on how to do this were placed in newspapers: turn off air conditioners when you aren't home, run clothes washers and dryers before 8 a.m. or after 6 p.m. if possible, and save the use of power tools for the weekend.

Economizing on power was nice, but the main item in the company's plan to meet the rising demand, as you might expect, was to build more plants, including several nuclear units along the Hudson River, like the one already operating at Indian Point. Other utilities around the country were also being scrutinized more carefully now by state regulators and by special-interest groups that seemed to put considerations for waterways or air quality or small animals above the imperative of supplying New York City with energy.

More energy would be needed. You couldn't deny that demand never went down. How could it? Once accustomed to a certain standard of living, society was not going to settle for less electricity. Some things were almost too obvious to be worth saying. As we were soon to learn, however, it ain't necessarily so.

While electricity planners in New York were beefing up their blueprints for meeting the increased demand, an event occurred thousands of miles away—an event having nothing to do with electricity—that almost overnight would cause grid history to take a great swerve. On Yom Kippur, the High Holy Day in the Jewish calendar, a barrage of gunfire erupted in the Sinai Desert on October 6, 1973, and war between Israel and neighboring Arab nations broke out on several fronts.

TOO MUCH GRID

If westerners didn't know much about the Organization of Petroleum Exporting Countries (OPEC) before, they would now learn quickly. OPEC had as members such nations as Venezuela and Nigeria, but its most potent bloc consisted of certain oil-rich Islamic lands of the Middle East, and officials there were bent on economic revenge for frustration on the battlefield. Shortly after the war began, an embargo on oil sales to the United States and the Netherlands was imposed in retaliation for their support of Israel.

The embargo lasted only five months, but its effect was profound in demonstrating the political muscle of the exporting states and the fragility of the world's energy supply. The cutoff and the subsequent huge oil price increases had an immediate effect on petroleum-related industries, such as automotives. A sideways impact was then felt by other heavy users of energy, such as the electrical business. Prices at the gas pump and at the kilowatt meter shot up.

Then the unthinkable happened: The growth in electric demand slowed. For the first time since World War II, the use of electricity in the United States for one year, 1974, was less than for the previous year. Many utilities, confident that demand would quickly return to its old gradient of ascent, stuck with their game plan. They pushed ahead with gigantic construction plans. In the 1960s they'd been embarrassed by a shortage of generation and weren't going to let that happen again. Even with higher-priced fuel, even with new environmental restrictions on smokestack emissions, even with escalating construction costs for nuclear reactors, the building boom continued.[7]

But demand did not go back up, or at least not at the old rates. One of those heavy ironies that characterize our tale now thudded into place. In the space of only a few years, in the 1970s and 1980s, the United States went from having a shortage of electricity-making capacity to having an uncomfortable surplus. Gradually it became apparent that some of the plants weren't necessary, especially the expensive new nuclear facilities. In a few cases there was talk in the back room about abandoning some of the plants.[8] It had come to that. Which was more painful, throwing good money after bad or declaring to your board of directors and investors that you had planned wrongly?

One hundred nuclear plants begun between 1974 and 1982 were cancelled, with billions of dollars in losses. Reactor construction costs had quadrupled, and only a small part of that was due to inflation.[9]

A new problem, a finance problem, loomed. Making electricity had always been a capital-intensive enterprise. That is, it had taken several dollars of infrastructure to make one dollar of product. Now it was getting worse. Bond ratings for some utilities dipped here and there. Con Edison failed to pay its shareholders a dividend in April 1974, a very cruel month, when the largest monthly utility stock decline since the 1930s occurred.[10] For those who operated the grid, the 1970s seemed like a deep pit with slippery sides and no apparent way out. It was a wacky world. In this winter of discontent electrical demand had stopped growing at the old rate, fuel was expensive, OPEC was ascendent, and inflation was moving up. Was there any good news?

Certainly not on the nuclear front, where the 1975 fire at the Brown's Ferry plant in Alabama (a reactor owned by TVA) and the core-melting accident at the Three Mile Island reactor in Pennsylvania in 1979 underscored the problematic nature of deriving electricity from splintering nuclei. No one was killed, but great psychological damage had been done. The Three Mile Island cleanup would prove to be more expensive than building the plant in the first place. The granting of nuclear licenses shrank down to nothing. The history of nuclear power, at least in the United States, had taken a swerve.

How the grid business had changed. Giant figures such as Thomas Edison or Samuel Insull were no longer to be seen. The generation and transmission of electrical energy, once a forefront engineering domain, now had all the excitement of accountancy. The distribution of voltage *was* a type of accountancy. Was it not the movement of a commodity in different denominations to various jurisdictions, followed by a bill at the end of the month?

The greater part of the power industry consisted of investor-owned utilities operating under monopoly conditions, which meant that only one electricity provider was available in most towns. This arrangement avoided a costly duplication of wires and poles and generation plants. And for the privilege of being the sole source of power, the utility submitted itself to the oversight of a state regulatory body that set the rates in such a way that the company made a profit but not a big profit. The whole system was sensible, dependable, and dull. In the power busienss, you could make a living but not a killing.

If there were little motivation for change, how would innovation ever come about? Altered circumstances—a war in the Middle East and high fuel prices—had imposed changes from outside the power industry. From the inside the realities of physics, in the form of thermodynamic laws curbing the efficiency of steam engines or the wayward behavior of atoms inside nuclear reactors, had imposed their own limitations on the way power was generated. Surely there must be some way that the managers and engineers of the grid could take fate back into their own hands and turn things around. But how? We thought electricity consumption would always grow. It didn't. Fuel prices had been reasonable for so long. In the 1970s the price of oil had more than quadrupled. We believed utility stocks were rock steady. They weren't.

Compare and contrast: In the computer business, "Moore's law" refers to the prediction made in 1965 by Intel official Gordon Moore that the computer power of a central processor—the number of transistors on a microchip, or roughly the amount of calculation performed per unit cost—would keep doubling every two years or so. This prediction has been (and at least for now continues to be) borne out, which helps explain the robust condition of the computer industry and Internet-related commerce.

Something like this had also been at work in the grid business, where through succeeding decades the amount of coal (or comparable fuel) needed to make a kilowatt-hour of electricity kept falling. This situation made electricity more and more of a bargain and helped amplify its already huge role in industrial and domestic life. It was nice while it lasted: ever-improving dynamos making electricity and central processors making calculations. But then Moore's law for the grid started sputtering out in the 1970s, and grid planners had to look elsewhere for new forms of economy. If ever Moore's law for semiconductor microcircuits peters out, the history of computers will probably take a swerve too.

Change in the grid business *would* come, slowly at first and then more quickly. Some of the changes were to come through engineering innovations, some through legislation, and some through business reorganization. In presenting the saga of these changes the remainder of this chapter will, for reasons of economy, reduce many complex events to a few representative actions, and the cast of characters will be limited. The deeds of decades will be telescoped to a few dozen pages. What you're getting here is not a comprehensive or scholarly *history* of the electrical grid but rather the essence of the *story* of the grid during some very turbulent years.

GREENER GRID

Foreign Affairs magazine, not exactly a primary outlet for news and views about the electrical grid, carried an article in its October 1976 issue called "Energy Strategy: The Road Not Taken?" The author was a young physicist named Amory B. Lovins, and he was eager to contrast two rival scenarios for the way society uses its resources and how business and engineering circumstances shape the petroleum and electrical industries.

Technologically advanced nations were following a "hard path," Lovins argued. In this scenario, primary energy originates mostly in immense oil and coal fields. After trans-shipment, the raw fuel is processed in massive refineries and generating plants (the bigger the better) and dispatched into extensive sophisticated networks of pipes and transmission lines. The metabolic result of this activity produces both desirable consumables in the form of electricity and numerous petroleum products and many undesirable byproducts in the form of noxious greenhouse-effect and acid-rain-forming gases or radioactive residues that would, if spilled, become a nasty contaminant of farms and cities.

Compare this, Lovins said, to a "soft path" approach to energy, one emphasizing efficient end uses and employing smaller-scale, more decentralized but no less efficient production and distribution. Fuelling this system, whether for transportation or for making electricity, would involve the gradual phasing out of fossil or fissile materials and the phasing in of renewable sources such as solar and wind power. This was not a new idea. In the mid-1970s solar cells had already been around for some time, and windmills had of course been used for centuries. Unfortunately, the direct use of sun and wind was still too expensive to deploy on a large scale for power production. The machines were unproven, couldn't provide energy in large quantities, and didn't seem to be compatible with the existing grid. Lovins's quest to chase after windmills seemed quixotic and was easy to dismiss.

Why was Lovins's energy outline published in a magazine devoted to foreign affairs? For one thing, the Arab–Israeli war and the subsequent oil embargo were then only a few years in the past. Moreover, energy was playing an ever more strategic role in all the immense issues on the world stage: economic development, climate change, health, and security. Lovins's essay was certainly not the first or longest assessment of energy or sustainable growth, but it received plenty of attention. It was a finely argued examination of the complex technological conundrum of how we can match rising world population with available resources, and it arrived just when anxiety in the West over access to energy supplies was quickly mounting. This essay was also the opening chapter in Lovins's prominent career as gadfly and persistent reformer. The *Foreign Affairs* article spoke plainly of the problems of the energy system then in place and how it could be improved in a major way.

The greatest compliment to Lovins's seriousness was the quick and pervasive chorus of praise and detraction that ensued, especially the latter. If Lovins emphasized the efficiency and ideal features of the soft path—a call for using only a fraction of the energy then being consumed—then that is what his critics seized on; this was not a soft path but a soft-headed path. It would guarantee a lower standard of living, critics said. It was a return to the simpler life of the Dark Ages.[11]

Some skeptics made the mistake of believing Lovins was some mere hippie, a countercultural complainer. Counter he might have been but not a complainer. He was not Tiresias or Jeremiah come to prophesy doom. He had suggestions, lots of them, and he wanted to put them into practice. Lovins would keep pushing the great idea of energy efficiency. Again, the principles of energy conservation and energy efficiency were not new. What Lovins did was package the ideas by presenting them in book after book, and lecture after lecture. He sought to get the ear of government officials, businessmen, and trade groups. It helped too that he was generally better informed than his critics.

What was the heart of his plan? To start with, he argued, the biggest source of *new energy* was right in front of us, not buried in the ground needing to be mined but hiding in our own homes and factories in the form of wasted energy. Begin with 100 units of primary energy—coal, say. In generating electricity in a typical power plant, 66 units of energy go right up into the sky as waste heat. For automobiles the waste is even worse. It's as if, when you cooked a meal, you were to take two-thirds of the food and immediately dump it into the garbage.

Then, when we *use* the electricity, there is further waste, as when inside a lightbulb heat is made along with light. To continue with the meal analogy, it's as if, after you served the food (the one-third that was left), each diner threw away 90 percent of that. Wouldn't that be appalling? You'd have to say there was something wrong with any food plan that operated this way. In other words, about 3 percent of the energy in the fuel gets turned into light.

Many things can be done right away to reduce energy waste, especially if you have a personal stake in the matter. For example, on a freezing day you put on a coat. You don't think to do this because it is morally right or politically correct or energetically efficient. You put the coat on to stay warm. The coat keeps the cold out by keeping the warmth in. That warmth comes from food energy, and on a cold day you don't want to waste any of it, so you instinctively insulate a warm region (your body, at an internal temperature of 98.6 degrees Fahrenheit) from a cold region (the snowstorm all around you, at a temperature of 23 degrees). You can also put a coat on your home in the form of additional insulation in the roof, walls, even the windows. On a cold day you wouldn't leave your coat unzipped, so why leave your home unzipped?

The general concept of energy is pretty direct. You put food into your mouth, gasoline into your auto, or electricity into a lamp. The energy is either visible to the eye or palpable to the mind. So is the result. You move; your auto moves; the lamp emits light. Saving energy seems like a simple idea, but it is rather more difficult to visualize. What does "saved" energy look like? Lovins devised the idea of "negawatts."

If, he says, you replace an ordinary 100-watt incandescent bulb with a compact fluorescent bulb that casts the same amount of light and fits in the same socket but only draws 20 watts of power, then you have done something extraordinary. By *not* using 80 of the 100 watts, the new bulb is in effect a *producer* of 80 watts. It's as if your lamp were a tiny electrical generator sending 80 watts of power *back* to the grid. The 80 watts *not* used in that lamp are free for use in some other appliance or some other home. Or maybe they didn't need to be dispatched from the power plant in the first place. An equivalent amount of coal *wasn't* burned in any furnace. A small puff of polluting gas *didn't* go up the flue. True, by not using that small increment of electricity, the utility's revenue declined by a tiny amount, but then too it didn't have to make that much electricity, so the cost of doing business just went down a tiny amount. The customer consumed less, the utility produced less, and the atmosphere was less burdened with carbon dioxide.

A negawatt is an absence but a commodity nonetheless, one that has value. To create this value requires investment. You have to buy the efficient bulbs, which are more expensive than ordinary bulbs, and you have to buy extra insulation for your ceiling. However, it is usually the case that these investments pay you back through energy saving. And thus subtly do we assemble, one bulb at a time, home by home, the equivalent of a power plant, a plant we displace through thrift, a plant we can do without.

Buying efficiency is often cheaper than buying a new power plant. This is one of Lovins's cardinal principles: *Saving* electricity is cheaper than *making* it. (This might have been one reason Con Edison started its "Save a Watt" campaign during the brownout years; in the company's dire need to meet a growing power demand, it was cheaper and quicker in the short run to have its customers use less electricity than it was to bring new power plants into service.) Ben Franklin's adage, "A penny saved is a penny earned," applies directly to watts.

Suppose we do that—buy all those bulbs, and save all that energy. Isn't that the end of the story? No, it isn't. Saving energy on lighting is just the beginning. Energy efficiency is not just a short effort, a one-time purchase of a better product. The first step in exploiting energy efficiency is to see it as a rich source of energy, a mine from which still greater riches could be hauled. The second step is to recognize that this mine is quite deep and that we have only extracted the uppermost layer.

It might have been a season of discontent for the power industry, but it was a golden age for energy efficiency. From 1977 to 1985, the U.S. gross domestic product (GDP)—that handy, single-number register of the nation's economy—grew by 27 percent even as net oil imports fell by 50 percent, and net imports from the Persian Gulf were down by 87 percent.[12] This lowering of imports was, unfortunately, to be reversed years later, but for a brief time the notion of energy reform took on a substantive form.

Consuming more energy usually means greater national prosperity, so countries with large economies usually have large per-capita energy consumption. Poor countries use less energy. Beyond a certain point, however, this correlation needs qualification. The really sophisticated societies are those that make clever use of energy. Their prosperity rises even as they use *less* electricity than they did before to perform the same task. This cleverness is embodied in a parameter, the *electric intensity*, defined as the ratio of GDP to per-capita electric use. If a country is being smart in its use of machines, its economy will keep going up even as its electric intensity goes down.

A distinction needs to be made between energy conservation—using less energy—and energy efficiency—getting the same or better service using less energy. Promoting efficiency is not a program for deprivation. It does not mean a lower standard of living. Furthermore, the whole process of efficiency engineering, whether built into a system in the first place (a home with lots of insulation or an office where lights turn themselves off if no one is in the room) or retrofitted afterward as new equipment becomes available, can be profitable. Efficiency is not castor oil. It is not sacrificial or charitable in nature. An efficient machine saves resources, saves on repairs, and saves money.

Refrigerators and lightbulbs, requiring only a fraction of the electricity used by counterparts a few decades earlier, are getting better still. Since the 1970s, efficiency measures have saved U.S. consumers about $100 billion per year (for all types of energy, not just electricity). Lovins estimates that this is just the beginning; another $300 billion could be saved annually through further efforts. In other words, in our exploration for negawatts—saved energy or avoided energy—we have already located rich deposits,virtual Alaskas, and there are reasons for supposing that far bigger reserves are yet to be tapped.

Sure, air conditioners and refrigerators and light bulbs are much more efficient now, but don't we all use *more* power-grubbing gadgets than before? Don't computers represent a big new sink for sucking in electricity? Actually, just the opposite. If anything, the aggregate effect of the Internet, computer inventory, and computer control of other power-consuming machines—as part of our transformation toward a digital economy—is to *lower* the electric intensity of America. That is, computerizing commerce reduces the amount of electricity needed to produce the same amount of GDP as before. A scientist at the Lawrence Berkeley National Lab has determined that computer equipment accounts for only about 2 percent of electric consumption in the United States.[13]

Consider the problem at the architectural level, where vendors are often paid not on the basis of how low the cost can be for operating a house or factory, year after year, but by how low they can make the initial cost of construction. One of Lovins's favorite examples of "whole-system engineering" concerns the purchase, installation, and use of a pump for sending a fluid through a pipe. Usually the pump is planned and installed first. Then come the pipes which, to fill the remaining available space, must often be narrow and made to follow a bent trajectory. This adds a greater electrical burden, however, since pumping fluids through skinny angled pipes is much harder than through thick straight pipes. Planning the pipes first, says Lovins, and paying for the more expensive fatter pipes, permits buying a smaller pump, requiring less power. The less expensive smaller pump more than pays for the costlier fatter pipes *and* there would be a huge savings over the long run since the power required to run the whole system is less.

What to do about electricity—its cost, its pollution, its management—would take time and experimentation. In the late 1970s there was with President Jimmy Carter a willingness as there had been with Franklin Roosevelt in the 1930s to try something new in energy policy. FDR's main achievements had been, first, to curb the abuses of the utility holding companies and, second, to launch the Tennessee Valley Authority. Carter's main energy achievements were to spur innovation in automobiles, leading to much greater gas mileage, and to secure the passage in 1978 of sweeping energy legislation, including the Public Utility Regulatory Policy Act (PURPA).

The ostensible goal of PURPA was to obtain more electricity without using more fuel. It did this by fostering the existence of fledgling nonutility companies devoted to producing power through solar cells or windmills, or the simultaneous production of electricity and heat. In this cogeneration scheme the steam that has gone through a turbine to make electricity isn't just thrown away but is used to heat homes or drive industrial processes. This was an old idea. In fact, if you go far enough back in the history of electricity, back to 1900 or so, combined heat and power was an important way of doing business. The fraction of the energy in a shovelful of coal turned into usable power or heat could be as high as 80 percent. But as power utilities became more focused on making electricity and as the size and locations of the generators made it less convenient to sell steam, the cogeneration scheme went out of fashion. Utilities fell into the astonishingly wasteful habit of throwing away two-thirds of their energy as heat. Ironical that many plants were more energy efficient in the year 1900 than in the year 1980.

The cogeneration companies that remained were likely to prize the heat more than the electricity, rather than the other way around. It would be understandable if these companies wanted to sell their surplus electricity to the main grid utility in the area. The new legislation ensured this. Utilties now had to buy electricity from the independent producers (including factories with surplus electricity), providing the cost was lower than the cost it took the utility to make power for itself.

The utilities were not thrilled. Their business was making and selling electricity, not buying it from other companies. Furthermore, since the *scheduling* of electricity—the perpetual balancing act between load and generation—is a tricky thing, it would be an imposition to have to buy orphan power in small amounts and at odd hours. The utilities didn't like being forced to accept this mongrel electricity, and they fought the whole thing in the courts for years.

Eventually the independent-producer law was upheld. Here in the making was one of the great ironies of grid history. Here was a piece of social engineering, a law passed by Congress, aimed at wringing out a bit more efficiency from the power production process. In this it succeeded but with a gigantic unintended consequence; what had started out as only a small provision of the 1978 energy legislation would become a lever that would pry open the entire century-old electricity business.

UNBUNDLING THE GRID

The 1970s were messy for the power industry, and the 1980s looked to be more of the same. Further irony: There had been a power plant shortage, resulting in a building boom, resulting in a glut of power plants. The prospect of nuclear energy, which once seemed so promising, was souring. Disturbing accidents, massive construction cost overruns, and bitter licensing disputes began to lend a sinister edge to anything atomic.

The grid was stuck in a rut. To see why, look at the fundamentals. The monopoly status of most big U.S. utilities, a historic fact extending back to the early 20th century, didn't exactly motivate creative management. With a modest but guaranteed profit built into the business model, what incentive was there for changing the system? A utility's profit could not fall below a certain level. That was reassuring. But also it would not rise above that level. Where was the thrill?

The best engineering students, as if smelling the complacency, seldom went into the power business anymore. Computer science was where the action was. There initiative was rewarded, new products appeared often, and R&D was moving up, not down. There was a sense with computers, as there had been with electricity in Edison's day, that you could invent something that would have a positive impact on millions of people: the advent of personal computers, the growth of software, and the establishment of what would become electronic mail and the Internet. In addition—and this was hardly a drawback—you just might make a great deal of money in the process of changing the world.

Could the electricity business change? Could it be exciting again? Although utilities didn't necessarily want to give up their monopoly status (it's nice having the market to yourself), most could see that regulation by the state was both protective and encumbering. This might have been one reason Edison and Westinghouse, having launched their respective direct current and alternating current utilities in so many cities—the national and American leagues of electricity—decided to sell off their operating companies in order to concentrate on the inventing side of the business.[14] Wasn't it more fun to be a maverick, going where you please and charging whatever you wanted for a product—subject to customer demand, of course—than to be king of the hill and rule placidly over some single city under the tiresome guidance of a state regulatory body? Could a time ever come again when you could strike it rich selling electricity?

Then it came: the first sign of glacial melting. To see how evolutionary change came to the electrical industry, we must first look at changes in the natural gas industry. Also previously hemmed in by regulations, the capture of gas from underground reservoirs and its transport in vast pipeline networks across the country were being overhauled. Price caps were being removed, and new exploration had led to a much greater supply and to lower prices. And where was this fuel going? Into a new kind of electricity generator.

Actually, it wasn't really new, but it was newly economical. In a gas turbine the fuel is mixed with air and encouraged to explode. The force of this controlled combustion spins a turbine, which in turn activates an electrical generator. Sometimes converted from mass-produced jet airplane engines, the new gas turbines favored by the small start-up companies were still a bit expensive for the amount of power you got out (gas had come down in price but was still pricier than coal). However, they could be rapidly assembled, were relatively compact, and could be fired up quickly. Therefore, they were often used as a backup or during hours of peak demand. Even better, if you captured the escaping hot gases made in the combustion and used them to make steam and sent the steam into a second turbine, you could crank out even more electricity. With the old once-through process your energy efficiency was 30 to 40 percent. With the newer combined-cycle operation, using both gas and steam turbines, your efficiency could go up to 50 percent or more. Last but not least was the fact that gas is a lean fuel. Natural gas is essentially methane—molecules made of one carbon atom and four hydrogen atoms—compared to coal, which has much more carbon than hydrogen. Generally hydrogen is what you want. Combined with oxygen it provides the explosive thrust that turns turbines. Generally carbon is what you don't want. Combined with oxygen it makes undesirable carbon dioxide.

Here was something new in the electricity business. A small company, an unregulated company, could build modest generators and find a niche market. A gas turbine could dispatch power at peak times and sell it, under contract, to the local utility. Here was an opportunity for an energy entrepreneur to fit into the cracks between existing monopoly grid giants. This was a real incentive. If the independent producer could reduce its costs while supplying the contracted power, its profits would rise. This wasn't necessarily true for the utilities themselves, whose profit margin was pegged at a fixed amount above expenses.

Not only was the profit mandated but also the obligation to have an extra cushion of electricity-generating capacity at the ready in the form of numerous dynamos. The upstart independents could often make power more cheaply with their new machines, and the utilities had to buy that power. But what about all those older, less efficient dynamos the utilities had built years or decades before? In many cases these machines were still being paid for. These extra "stranded" costs had to be covered.

The utilities could say, "It's fine for the independent producer to do whatever it wants. It doesn't have the solemn responsibility, as we do, of delivering power to all customers all the time." But the law was the law. The independent producer could make power and the utility had to buy it. Some of the independent producers, or "merchant generators" as they were sometimes called, started to grow bigger and bolder. They now asked, "If we can sell power to the local utility, sending electricity over its wires, can we use those same wires to ship power to a different utility in a neighboring state?"

The U.S. Congress, in the Energy Policy Act of 1992, provided an answer. Yes, you may. Here was another new thing in the electricity business: long-distance wholesale shipment of power involving multiple interested parties. Now the independent producer had even more incentive to streamline its operations. It could sell to the highest bidder, even if that meant bypassing the nearby utility in order to deliver, or "wheel," power to a distant grid. Not only was the local utility left out, but its own transmission lines might be used to convey the current from the independent producer to the third-party buyer.

Yet another new thing in the electricity business: The company that sold the power might not be an independent producer, or a utility, but a broker. Without a single generator to its name, an energy trader could buy big blocks of wholesale power in one place and sell it in another. The most famous trader, Enron, didn't always sell at low rates, but they did offer to sell commodities other than electricity, such as insurance against future fluctuations in the price of electricity. Some utilities, now forced or seduced into buying lots of independent power, paid higher prices to ensure long-term stability of delivery. Enron offered this. It grew and grew. It came to sell as much electricity as any other company. It was the biggest trader of gas in the United States. It bought and sold in many commodity exchanges, and undertook colossal projects in developing nations. It referred to itself as the world's greatest energy company.

Fresh players had been allowed into the game. Unregulated producers of power could sell wherever they wanted. Energy traders, companies that didn't even make power, could sell it. In the space of only a decade or so the electricity business had spun around. Many utilities were still strong, but they no longer exercised total sovereignty. As regulated monopolies and long spared the rigors of competition, they now had to go on a crash diet. They trimmed their payrolls and started to sell off their own generators.

Why sell their dynamos? Because the utilities were finding themselves in an awkward situation: Part of the corporation (the generation part) was operating in an increasingly unregulated market. Meanwhile, other parts of the corporation (the distribution part, say) were still strictly regulated: The government decided how much customers could be charged. For the utilities it was sometimes easier to just get rid of their unregulated departments. However, if the utilities sold their generators, they put themselves in the odd position of becoming customers themselves, customers of those same independent producers they had fought off years before. In some states, utilities were actually required to sell their generators.

A great transformation was under way. It's as if the Ice Age were ending, and the emergence of a new climate were altering the landscape. Some of the old superpowers of the electricity world were unbundling their functions. What had been vertically integrated enterprises—some of them incorporated in the 1880s—were now being disestablished layer by layer in the 1990s. As the nature, size, and look of utilities changed, company names changed too. In the very earliest days of the grid (1880–1900) the word *illumination* would have been prominent (example: Edison Electric Illuminating Company), befitting the principal responsibility of supplying electric current to filament bulbs. Later, as effective electric motors became available (1900–1920), the firm's name was altered to incorporate both light and power. Then as the power part of consumption grew to be the majority product, the firm was often referred to simply as the "power company." Later still, in the age of diversification and trading and deregulation, and as the corporate mission grew more vague, so did the name on the letterhead. The fashion was for names that evoked the idea of energy without being too specific about it. Examples include Cinergy, Dynergy, and Entergy.[15]

A great fuzziness had settled in around the concept of monopoly grid management. The wires that fed electricity into your home were owned by the local utility, whose operations were still regulated, meaning that the rate you paid each month was set not by market forces but by a set of commissioners sitting at a table in the state capital. Retail rates were fixed, and yet that same power, a millisecond before it reached your house, might have been subjected to some unregulated, computerized, wholesale negotiation among companies hundreds of miles away.

Then people started asking, "Why not go all the way and deregulate retail electricity?" In a number of countries, such as Britain, Chile, Norway, and Australia, reducing controls on retail pricing for electricity had been tried—with plentiful complications, to be sure, but also some success. In many of these cases, the power industry had previously been run entirely by the government, so the transition to denationalized/deregulated electricity would have been even more of a contrast. So why couldn't it work in America, home of free markets and high octane competition?

Besides, a number of other industries had been deregulated and, as entrepreneurs would say, liberated. They had been turned loose to try new ideas and offer new products. Here are some prominent examples. Deregulating the natural gas industry had encouraged investment, and this led to new exploration and new supplies and, among other things, to the craze for gas-turbine electric generators. Deregulating the airline industry, after some initial heartache and bankruptcy, generally resulted in lower prices and more services. Deregulating the phone industry generally led to lower prices and more services. By contrast, deregulating the financial services industry in the United States led to the savings-and-loan disaster, in which many small savings institutions, suddenly free to invest in more lucrative but also more risky ventures, slid into an immense pit of default.

Would deregulation work in the power business? Some observers felt that competition at the wholesale level—selling large blocks of power—had been a useful innovation but that competition at the retail level was likely to produce no further economy, only chaos. Nevertheless, for many business leaders and capitalist-minded legislators, the lure of free markets was worth that risk. The gung-ho argument is simple: Get government rules out of the way and thrift and innovation will follow. Inefficiency would be penalized and ingenuity rewarded. Lower prices and more services would result.

The go-slow faction was more skeptical. Would markets be as free as advertised? Just because 10 power companies were supposedly competing doesn't necessarily mean lower prices would result. Just because consumers would have more energy choices doesn't mean they would choose wisely. Wouldn't all the large electricity consumers like factories, shopping malls, and chain restaurants, work out their own electricity deals, getting low rates, while ordinary customers would face uncertainty? The traditionalists cautioned that maybe it would be better to leave things alone. Electricity service wasn't getting much better, but it wasn't getting worse. Play it safe.

Inertia, personal and institutional, is indeed a considerable force in human affairs and a brake on possible change. The desire to improve things, however, is also a powerful force. Gradually, in the argument over what to do with the electricity business, the scales were tipping toward deregulation. Most people were pretty satisfied with their deregulated phones and deregulated airplane tickets. Why not deregulated electricity?

The federal government, with sway over energy that zips across state borders, had already moved wholesale power, or at least the generation part, in the direction of deregulation, thanks to those laws of 1978 and 1992. The distribution part, the delivery of voltage the last mile or so to the home or factory, was a matter for the states to handle one by one. The Federal Energy Regulatory Commission (FERC), the successor agency to the old Federal Power Commission, readily admitted the states' sovereignty in local cases but, just to be sure the overall deregulation message was getting through, it pressed utilities to sell off their power plants and their transmission lines, or at least to segregate those parts of their business from the distribution part. Why do this? Because the transmission lines were now considered a quasi-public conveyance to be used by the utility that owned the lines *and* by other makers of power, and it would be unfair (although understandable) if the utility favored power originating in its own generators.

To look at the overhaul of retail electricity, one has to go to the states, and the prime example is California. Not only does California have one of the largest economies in the world, making it much like a country, but its determined effort to reform the electricity business was to be the most conspicuous and illustrative case history in the short annals of restructuring in the United States up to this time. Californians are stingy with electricity. Their per-capita use is only about half the national average. Neatly neutralizing this advantage unfortunately is the fact that their electric rates are among the highest in the land, hence the desire to turn things around.

The California legislature, in consultation with the designated state body in charge of power, the California Public Utility Commission, decided in 1996 that controls over power rates would be phased out over four years. If the whole marketplace machinery could turn freely, the thinking went, the competing companies would become ever more effective and service oriented. Inefficient companies would lose the race and disappear. Enterprising companies would thrive and expand. Consumers would win too. If deregulated air transportation meant lower prices and deregulated telephones meant lower prices, shouldn't everyone get lower-priced electricity too?

CRASH LANDING THE GRID

Early in 1998 the great deregulation experiment began. Wholesale power was now to be sold at "market prices." The California Power Exchange opened for business with a lineup of wholesale producers offering to sell power to wholesale traders and distributors. At the end of this first day the price of electricity had actually dropped a bit, just as you would expect.[16] As for retail rates, because of the way utilities were recovering their past investments, price caps came off in stages. At first the deregulation settled on customers living in the San Diego area. Residents there would be "free" to order electricity from several competing companies. The old utility, which owned the wires going into homes, would receive a carrying fee but wouldn't necessarily be providing the energy. The arrangement was not unlike modern telephone service.

Rates stayed steady for about two years and then started to climb. There now followed a cause-and-effect roller coaster ride of pricing events. Customers who stayed with the utility screamed at seeing the higher prices, so the state eventually reinstated caps on retail rates in San Diego. To counter this measure, the independent power-producing companies started to sell power to buyers in a neighboring state who would then sell it back to buyers in California at a higher price. In this way California price controls could be avoided. In response, one arm of the federal government, the Department of Energy (DOE), used emergency powers to force producers to sell more electricity in California, while another arm, FERC, removed the state-imposed caps once again.

By this point wholesale prices were so high—not just in California but all over the western states—that some aluminum manufacturers in the Northwest discovered that they could make more money by suspending their metallurgy and going into the power business. They figured that, instead of operating the power-intensive smelting machines, it was more profitable to turn them off and sell the unused power, bought on contract at low prices, in the open market for high prices.[17]

The crisis reached its peak in the early months of 2001. Two of the largest utilities in California, Pacific Gas and Electric (PG&E) and Southern California Edison, were caught in a deadly bind. They were forced to pay five times more for power (in a deregulated wholesale market) than they were allowed to charge their customers (in the still partially regulated retail market). These companies had to borrow money to buy power, which they then sold at a loss. Every day they remained in business they went millions of dollars further into the hole. The utilities' credit ratings fell, and the independent producers, fearing they wouldn't be paid, started balking at making deliveries. In some areas on some days there were rolling blackouts, the first since the Second World War. Governor Gray Davis declared a state of emergency, and the state itself started to buy power on behalf of the beleaguered utilities. PG&E declared bankruptcy anyway.

Retreat was the better part of deregulatory valor. By autumn, for the time being at least, deregulation was suspended, and the price of electricity stabilized again at levels not so different from where they'd started. The immediate threat of power shortage was over. The spectacle, while it lasted, had drawn a huge audience. Many other U.S. states and several governments abroad—anxious onlookers of the great California experiment—now rethought their own plans for deregulation. Some states wend ahead, while others held off.

Affixing blame for the electricity fiasco now intensified. Many residents, not wanting a generator in their backyards, had been happy to see power shipped in from elsewhere, which is fine providing that the residents of those other states aren't having a power shortage of their own. California wanted the benefits of electricity but didn't want unsightly power plants, it was said. It didn't want risky nuclear plants built near seismically active faults ("quake and bake"), and didn't want pollution pouring out of smokestacks. Other factors: continued drought in the West meant that rivers were low and therefore hydroelectric generation was diminished accordingly; natural gas, which was now fueling more and more of those fancy new turbines, suddenly got much more expensive.

No, the problem was greed, others argued. Independent power producers and traders such as Enron and Dynergy had manipulated prices higher and higher. They discovered that they could make more money by holding back power, selling less of it but at vastly inflated prices. They had "gamed" the markets. That is, they had applied the mathematical rules governing game theory to direct their sales and purchases, to determine when to withhold and when to come forth, so as to maximize profit. Lost completely was the notion that delivering power was a civic service and an important obligation for the public good. As Californians had suffered through blackouts, the power producers' profits had in some cases gone up more than 100 percent in the year 2000.

We're in business to make money, the companies would respond. Maximize profits? What else are we supposed to do? We played by the rules. If you don't like the results, then change the rules.

The people who had made the rules, the legislature and the utility commissioners—it was their fault, some said. They had shortsightedly forbidden the use of long-term contracts. Power was to be bought wholesale by utilities at very short intervals in the spot market, the better to get the very lowest prices at that moment. The thinking was that prices would indeed be low since all the producers would be competing, gladiator-like, in the energy arena. Sellers struggle for existence, while buyers get to pick and choose. This is capitalism at its finest.

But it ain't necessarily so. For one thing, the number of wrestlers in the ring—the independent producers—was never that big. For another thing, the lack of long-term contracts lent a perpetual instability to the whole enterprise of buying and delivering power to millions of customers. This wasn't capitalism, some critics said. The markets weren't free. Some parts of the business were deregulated, while other parts remained regulated. You can't just take the caps off wholesale prices. What about retail? So don't call it "deregulation." Furthermore, and most obvious of all to anyone who has ever invested in the stock market, prices can go up as well as down. Many ordinary customers, and state legislators too, seemed to have thought that electricity prices could only go down and were surprised when they went up.

Then, at the worst moments of the energy shortage and price crisis, the utilities had been stuck in the middle, having had to supply power to customers at regulated low prices and, at the same time, buy power from independent producers at high unregulated prices. What could they have done? Once masters of their own fates (since they made their own electricity) and pillars of public life, the light and power companies now were fighting for survival.

There were plenty of losers in this adventure, starting with customers suddenly forced to absorb huge rate increases. State regulators criticized federal regulators and vice versa. The feds, eager for increased competition, were inclined to cut the markets loose from all controls and let the going rate find its natural position, whatever it might be, until the rate ascended into the heavens. They later admitted that they should have intervened.[18]

Irony overflows. Instead of new services, there were blackouts. Some of the large energy companies, which had made a fortune at the height of the crisis, were themselves struggling near the edge of bankruptcy a year or two later. Finally, Governor Gray Davis, who had nourished hopes of living in the White House, couldn't even hang onto the state house. He was turned out of office in a recall referendum, in large part because of his handling of the electrical mess. He was replaced by a movie star.

So who was to blame? A factor cited by many commentators was the sluggish growth of generating capacity. The California economy was getting bigger faster than the inventory of dynamos. Should it be any wonder then that power shortages should develop, whether or not prices were deregulated? On the other hand, as Amory Lovins points out, in the summer of 1999 the California grid successfully handled a peak load of 53 billion watts, whereas in the crisis moment of January 2001 a peak load of only 29 billion watts was enough to trigger partial blackouts.[19] About this time 10 billion to 15 billion watts of generating capacity were "calling in sick." That is, the plants were down for repairs when they were needed the most. The power companies answered: We needed to do routine maintenance on our machines. You can't fault us for that.

Or can you? The California regulators issued a report which asserted that many service interruptions could have been avoided if the big independent power companies operating in the state had kept available generating capacity in use.[20] Later, federal regulators also asked several of the energy companies to document their actions during the months of acute power shortages.[21]

Numerous lawsuits against the power producers were filed to recover some of the heavy payments for electricity made during the crisis. The plaintiffs charged that the companies artificially contrived the energy shortages in order to jack up prices. In one of the proceedings, a tape recording was released of a conversation held between an Enron trader, identified as Bill Williams, and a Nevada energy company official, identified as Rich. Here, from the pages of the *New York Times*, is a brief transcript of the conversation between Bill and Rich. It took place a day before the power plant in question was taken out of service on the very day of a rolling blackout in California:

> "This is going to be a word-of-mouth kind of thing," Mr. Williams says on the tape. "We want you guys to get a little creative and come up with a reason to go down." After agreeing to take the plant down, the Nevada official questioned the reason. "O.K., so we're just coming down for some maintenance, like a forced outage type of thing?" Rich asks. "And that's cool?"
>
> "Hopefully," Mr. Williams says, before they both laugh.[22]

And what about Enron? The lawsuits began immediately. Enron, the suits claimed, had launched weird projects, with code names like "Death Star," for making quick profits even if it meant bringing down the grid. The company exported power out of state only to bring it right back in ("megawatt laundering") in order to avoid price caps.[23] But Enron was Enron and thought they could do anything. Its profit-making potential had impressed many investors. The impressiveness was apparently helped along by fraudulent accounting practices (most of which, to be sure, had little to do with the California mess). Declared by *Fortune* magazine as "America's Most Innovative Company" several years running, Enron's own investments and projects came to grief, and the company was forced to declare bankruptcy in 2001. The company's demise was the largest business failure of any kind since the Insull crash in the 1930s. Enron's questionable bookkeeping scheme soon came to light, and criminal convictions of top company officials followed.

WHAT KIND OF GRID?

One of the main reasons for the California grid turbulence of 2000–2001 was the stormy marketplace. Robert Kuttner, who writes frequently about economic matters, says that a true market brings together willing customers and willing buyers around a price mechanism that "functions to apportion economic forces efficiently; they signal sellers what to produce, consumers how to buy, capitalists to invest."[24] But, Kuttner argues, capitalism in practice doesn't work this way. Instead what we have is a mixed economy:

> The idea was that market forces could do many things well—but not everything. Government intervened to promote development, to temper the market's distributive extremes, to counteract its unfortunate tendency to boom-or-bust, to remedy its myopic failure to invest too little in public goods, and to invest too much in processes that harmed the human and natural environment.[25]

In a true democracy we have the principle of one man–one vote. In true capitalism we have the principle of one dollar–one vote. But in the great arena of human affairs, this ain't necessarily so. Affluence buys influence. Few social institutions are truly true. Freedom allows citizens to be virtuous or selfish. A competitive economy rewards innovation and ingenuity. It also rewards greed and manipulation. It's prudent to have in place some rules to encourage good behavior and discourage bad behavior. And who decides what's good and what's bad? Well, that's what social discourse and legislative action are for.

Has regulation been that bad? Kuttner asserts that many of the most dynamic industries of the past century in the United States—telecommunications, aviation, electric power—have been subject to mostly effective governance and that the independent regulatory bodies—such as the Federal Communications Commission, the Federal Aviation Administration, and the Federal Energy Regulatory Commission—constitute practically a fourth branch of national government.[26]

The redrawing of electric regulations hasn't exactly polarized into a Republican versus Democrat division, but it has become part of a larger debate over the economic rules of conduct for our Western capitalist society. How much risk do we take? With a resource as vital as electrical energy, shouldn't we play it safe? Or does that preclude innovation? How much money should we invest, and what return can be expected? At what point does enlightened self-interest become rapaciousness? How large a role should government play? Indeed, how big should government be?

Deregulation hits home. In my own corner of the grid, in southern Maryland just outside Washington, D.C., the newly deregulated monthly bill is about to go up almost 40 percent from what it was a year before. The explanations given are chiefly higher prices for the fuel used to make electricity and adverse fluctuations in the price of auctioned power. In this case the company delivering electricity to my home happened by ill luck to buy a year's consignment of power from the company making electricity just as the price shot up.

Where do we stand? The grid overhaul still under way was initiated by several forces. Among these were the arrival of new technology; fuel shortages brought on by instabilities in the Middle East; reactor accidents and ballooning nuclear construction costs; worries about energy efficiency and environmental degradation; and a succession of new laws that, deliberately or inadvertently, ushered in a large-scale realignment of an industry that previously had been chiefly a monopoly preserve of large utilities. The turbulence of the California experiment notwithstanding, many of the innovations have been welcome. In some U.S. states and some countries, a bit of reregulation has been necessary to correct various imbalances caused by the lack of true market competition. But restructuring goes forward. There is no going back.

Go back to what? The world of 1965 doesn't exist anymore. That's when the great November 9 blackout struck and when the decades-long Moore's law for electricity—the unit cost of making electricity going down through the years—came to an end. After that the grid, seeking to reinvent itself, swerved through some tumultuous decades. The grid of 1965 is gone.

Utilities have cut their budgets. Unfortunately, some customers would say, in the effort to stay competitive, their power providers have often scrimped on service. Repair trucks are slower to respond to problems. Power interruptions are more frequent. R&D for new equipment has declined. The construction of more transmission lines has not kept up with the demand for new electricity. The amount of surplus generating capacity—the standby power a company keeps for emergencies—has shrunk. The need to stay competitive means costs have to be kept down, the utilities say. They have to sail closer to the wind, closer to a condition of potential instability, and farther away from the grid of yesteryear.

Restructuring had altered the social contract. The old obligations weren't there anymore. The understanding, from Insull's time onward, was that the utility guaranteed service in return for guaranteed profit. The utility was a private company but also a very public fixture in society. Now the utility would be a vendor, one among many.

In this era of Enron-like machinations, a new vigilance is required. Federal legislation, such as the Sarbanes-Oxley Act of 2002, requires more detailed financial accounting procedures by corporations. And working up from the grassroots level, a movement toward socially responsible investing is becoming popular. It's not enough for a company to make a nice profit. Some investors insist that the company's products benefit customers, that employees are treated fairly, and that environmental impacts be taken into consideration. Rate of return *and* ethical behavior are expected.

There is no going back to the old electrical grid. Consider that before restructuring began in earnest in the mid-1990s, much of electric technology was basically a scaled-up version of the grid as it had been in the 1940s. Customers were getting pretty much the same electric services in 1995 that they'd been offered in 1945. What if, one critic asked, the airline industry had stagnated in the same way? Then modern aviation would look something like this:

> . . . top speeds limited to 200 mph, only short hops between adjacent cities, with transcontinental travel requiring many frequent stops, low altitude operations that make flight bumpy and cause frequent delays due to storms, and prices affordable only by the wealthy.[27]

8

ENERGIZING THE GRID

Vertigo. I don't suffer that much from fear of heights, but taking the fresh air here at the 12th floor is disconcerting. This gap in the wall, a sort of garage door 150 feet up, is where supplies are hoisted into the building. Standing in the gantry, I am separated from the sky by only a frail cord. Below, at the foot of what would be a hurtling few-second fatal fall, lies an immense switchyard where electricity is carved for the separate boroughs of New York City.

The sensory effect of vertigo, the trick of the mind in which we realize (in order to save our lives) the horrible mismatch between the human scale and the daunting height of the cliff in front of us, ought to apply to high voltage, but it doesn't. Electricity wasn't around in the long Stone Age when the mental faculties of our species were being tried out as we roamed the East African savannah. Therefore, we don't have a visceral dread of electrical heights that we do for geological heights or for spiders and snakes. It's hard to appreciate the magnitudes involved. You can see the energy manifest in Niagara Falls: There it is, all that water thundering downward. But how do you gain an intuitive feeling for the 345-kilovolt currents emerging from a modern major dynamo?

Perhaps you've seen a movie in which half a man's beachfront house has been washed away by a storm overnight, and as he opens the door expecting to walk into a bedroom he falls directly into the ocean. Or the advertisement for all-season auto coolant in which a woman, dressed in shorts inside a sweltering apartment, opens her door and strides into a driving snowstorm. This jarring mismatch of reality pretty well typifies the queerness of electric power. On one side of reality is that innocuous wall outlet with its modest slots silently waiting. On the other side, the grid side, is a kick almost too great to calibrate. Comparing your personal energy level, the strength in your muscles fueled by the food in your stomach, to the grid energy level is like comparing Rhode Island to the planet Jupiter or a paper clip to an aircraft carrier.

We have come now in our voyage around the grid to a place where a lot of that incommensurability comes about. We are at Ravenswood, the largest power plant inside the borders of New York City, and standing 10 floors above Big Allis, still after 40 years the greatest single generator in town. So far in this saga of electricity we've been *at* but not *in* any of the great citadels of power. We've stood outside long enough, and it's time to come indoors to inspect things up close.

THE HALL OF DYNAMOS

Ravenswood is a fenced-off compound in close proximity to teeming apartment buildings at the western fringe of the borough of Queens. The smokestacks are half as big as the Empire State Building across the East River, and the main building is one of the biggest things in a big city. You'd think it would be on the itinerary of more motor tours of the Big Apple, but it's a bit hard to get to if you're starting out in Manhattan.

After going through the security check, the first thing you see after coming in the front gate is the big scoop once used to lift coal, tons at a time, out of barges bobbing in the East River. To keep Allis fed required up to 2 million tons of coal per year. That changed in 1971 when her taste turned to oil.Today oil is delivered at riverside, but the fuel used inside is actually natural gas. Allis can adapt to gas or oil; in a world of uncertain fuel availability and price volatility it helps to be an omnivore.

A visit begins with a badge, a hardhat, and a background briefing. Here in the office, a digital readout immediately tells what's happening. Unit #10 is humming along at 300 megawatts, unit #20 is down for maintenance, and unit #30 is coasting at 510 megawatts. This is about half of what it's capable of doing; the rest is being kept in reserve.

Now we get down to business. Walking along several coldly lit hallways, up steel stairs, through a series of doors, we enter the vast turbine room. Numerous fluorescent fixtures hang about and yet, because of the immensity of the space to be lit, all the visible surfaces seem dim. We can hear but not see where steam is shot against the blades of a heavy pinwheel, turning a shaft, rotating one coil of wire past another coil, sending gusts of electric force into circuits waiting elsewhere. The flooring is made of the dimpled metal sheeting one sees on the deck of a battleship.

The main sensory impressions are of sound and warmth. The sound comes from a mixed chorus of motors and pumps, while the heat comes from the ceaseless burning of fuel. Electricity, not lava, is being produced; otherwise, you'd say you were inside a volcano. Unfortunately, all that warmth is wasted. Of all the chemical energy contained in the fuel consumed every moment only one-third is converted into electrical energy, while two-thirds merely makes New York hotter. It's those two-thirds we feel the whole time we're indoors at Ravenswood.

Unit #30 lurks at the far end of the hall, a sort of grimy white in color, looking like Moby Dick beached. Appearing a bit rusty, the company's nameplate, "Allis Chalmers," from which the unit gets its familiar name, is still affixed to the side. The rust and our host's cheerful nonchalance and the noise and grime do not lend poignancy to the moment, and yet I am feeling the kind of self-consciousness I get when visiting the high altar of a great cathedral. After all, this is Big Allis, the most storied still-active generator in America.

Here, just a few steps away, is where grid engineering culminates. This is the crux of the power industry. It all comes down to these two great complementary machines: the turbine—focal point of the steam cycle, where thermal energy is converted to mechanical rotation—and the generator—focal point of the electrical cycle, in which the rotation of a wire coil induces powerful currents zooming forth to power the city. In between turbine and generator, in a gap only a meter or so wide, we can see the common central shaft, blurred out by spinning at 3,600 revolutions per minute. This is the shaft that lacked the oil, whose bearings were burned, that spun to a halt, that finished the crash of the system that Edison built.

Today, more than 40 years after the 1965 blackout, it's alive with energy. Standing so close to this concentration of bottled lightning, furnishing power for maybe a million people, the visitor might be wondering how to behave at this moment in this place. The machine is impressive, and shouldn't we summon emotion proportional to this mighty force? If there is a mismatch between the expected feeling and the actual perception, perhaps this is because the visitor will have lived his whole life with electricity. It's been assimilated into the scaffolding of everydayness.

In search of an articulated sensibility, let's return to Henry Adams, the historian and wry observer and questioner of social and technological change. On one occasion he too had been brought to a hall of dynamos by a scholar friend. Adams stood very close to the spinning apparatus and had this impression:

> To him [the scholar], the dynamo itself was but an ingenious channel for conveying somewhere the heat latent in a few tons of poor coal hidden in a dirty engine-house carefully kept out of sight; but to Adams the dynamo became a symbol of infinity. As he grew accustomed to the great gallery of machines, he began to feel the forty-foot dynamos as a moral force, much as the early Christians felt the Cross. The planet itself seemed less impressive, in its old-fashioned, deliberate, annual or daily revolution, than this huge wheel, revolving within arm's length at some vertiginous speed. . . . Before the end, one began to pray to it; inherited instinct taught the natural expression of man before silent and infinite force. Among the thousands of symbols of ultimate energy, the dynamo was not so human as some, but it was the most expressive.[1]

Maybe Adams was being poetic in equating a huge dynamo with a moral force. Or he might have been sincere. Henry Thoreau, had he lived into the grid age, would probably have considered the dynamo superfluous, at least for his way of life. He would, however, have to admit that in later times electricity was the great facilitator of powered activity as steam had been in the 19th century. As for Lewis Mumford, he would have readily agreed that the dynamo was a moral force, meaning that it was a force, or at least a potent man-made instrument, for doing good and ill. With electricity you can energize machines for saving or taking lives. All three—Thoreau (c1850), Adams (c1900), and Mumford (c1965)—would recognize Allis, the ultimate electricity machine, as being profoundly expressive of its era.

DIVESTED ENERGY

Hidden beneath the Ravenswood turbine deck, below all that metal flooring, is more metal: insulated metal pipes carrying steam from furnace to turbine, metal pipes carrying cooling water from the East River into the plant, and the 12-inch-wide copper blocks, the bus bars, carrying the primary electricity out of the generator coils. If Big Allis is the heart of the New York City electrical system, these thick conductors constitute the aorta, the first and largest of arteries for conveying the precious energizing bloodstream through to other parts of the circulatory system.

After Genesis comes Exodus. The freshly made electricity shoots forth at 22,000 volts but is quickly transformed to an even more fabulous 345,000 volts for dispatching to the surrounding grid. New York real estate being notoriously expensive, Allis's circuit breakers—the main on/off switch for cutting off currents—are located on the roof. Consequently, the huge tide of electricity is immediately routed to the outside of the building and then upward. To my eye, these power lines, held away from the exterior wall by a large brace, look just like the strings of a musical instrument. That is, the Ravenswood building, viewed from the side, resembles an immense electrified cello.

Instead of producing sound waves, though, Ravenswood produces electric waves of a particular purity. Ravenswood's melody is always a monotone 60 cycles per second. If things go well, this continues 60 seconds per minute, 60 minutes per hour, 8,760 hours every non–leap year. If you multiply all those numbers, you get 1,892,160,000 cycles in a year. So good is the clocklike precision of spinning turbine motion in general that some generators, if they've been running without interruption for a year, will come within a cycle or two of this exact number. A few machines will hit it dead on.[2]

Ravenswood is no longer owned by Con Ed. The whole plant, along with Big Allis, was sold in 1999 to another company entirely as a side effect of the massive restructuring process that obliged or encouraged large utilities, the companies that actually deliver electricity to homes and factories, to divest themselves of some or all of their generating capacity. Many utilities now distribute electricity but don't make it themselves. With the stroke of a pen across a complicated legal document, Big Allis, the pride of Con Ed's energy lineup, suddenly found herself playing for a different team, KeySpan Corporation, one of the largest gas distributors in the country and the largest single generator of electricity in New York State. Restructuring has changed things. And yet things are also the same. New owner, old electricity. Under long-term contracts, Big Allis's output still goes where it has always gone, mostly to homes in and around New York City.

We are conducted around the building. One of the closed-circuit videos in the control room shows a continuous image of the firebox, the part of the furnace where the combustion takes place. This furnace, 12 stories tall, occupies the larger part of the building and certainly accounts for the pervasive heat felt everywhere. Even with extensive insulation all around, the temperature on one of the gantries we visited must have been 120 degrees Fahrenheit. The operations manager approaches a tiny viewing port in the side of the furnace, swivels it open, and the inferno within becomes visible: a swirling orange flame, a subvolume of the hydrocarbon hurricane within. Don't get too close, you're told. If for some reason the pressure were to be reversed, some of that glow, at a temperature of 3,000 degrees, would come shooting out at this outlet like a laser beam, incinerating anybody that had been spying through the keyhole.

What takes place behind that keyhole is of grave importance. Earlier I said that the focal point of the whole power enterprise was the linkup of the turbine and generator; from that spinning shaft comes the voltage that electrifies cities. If, however, we view things from a different perspective—from up in the sky or standing atop Canadian icefields or above vast tracts of ocean—our attention must shift to a different power couple: furnace and smokestack. What matters to the sky is not the electricity coming out of the plant through wires but the plume of tri-atomic molecules coming into the air.

In many countries a majority of electricity is made the Big Allis way, with fossil fuels going into the plant at one end and electricity coming out at the other end along with the undesirable byproducts. It can't go on like this. Leaving aside the argument over whether fossil fuels (coal, gas, oil) will be all used up in a few decades, there remain two major issues: efficiency and pollution. We can't afford to throw away two-thirds of all the chemical energy embedded in the fuel. Nor can the planet's inhabitants endlessly tolerate the devilish chemistry by which various atoms in the carboniferous fuel—such as sulfur and of course carbon—combine with air molecules to form compounds that make trees wither and humans wheeze and climate change beyond the natural variability always at work.[3]

ADVANCED CARBON

How can we make the same amount of electricity more efficiently and with less pollution? The ironic answer, over the short run, might be to burn even more fossil fuel. Not coal, but natural gas, essentially methane, whose combustion produces no sulfur, fewer nitrogen compounds, and far less carbon dioxide. How does it work? Just to the north of the Ravenswood's factory-like building is Big Allis's baby sister, the newest large generator to be built in the city in 10 years, a sleek, relatively low-emissions gas turbine, cleverly squeezed into a space of about 2 acres. It uses the latest noise abatement equipment, which makes all those nearby apartment dwellers happy, and has a high-efficiency, dual-turbine architecture. It won an award for being one of the top power plants for the year 2004.[4]

Gas turbines are quite popular at this moment in the history of the electrical grid. They can be built quickly and can be as large or small as a company wants, depending on the amount of extra electricity needed, from many hundred megawatts down to kilowatts. In a gas turbine pressurized gas is mixed with air to produce an explosion.

The hot escaping gas directly spins a turbine to generate electricity. The process is even more attractive when heat is captured from the exhaust gas coming out the back, heat used to make steam, which turns a second turbine to generate extra electricity. In such a "combined-cycle" approach—making electricity in a gas turbine *and* a steam turbine—the energy efficiency can go up to 55 percent. Efficiency isn't just a clever thing to do; it means using less fuel and producing less carbon, to produce the same amount of power. Result: more profit and less pollution. The utility comes out ahead; so does the customer, and so does the atmosphere. Still more efficiency can be wrung out of the whole operation. If the leftover steam is used for heating, the total energy efficiency (electricity plus heat divided by the energy input) for the combined heat-and-power process could be as high as 90 percent. Think again of the food passing through a restaurant kitchen. An able chef uses leftovers to the fullest. What can't be deployed in a salad can go into a soup or at least onto a compost heap. Only as a last resort should the scrap be thrown out. Why not do the same with energy?

Another way of rethinking the traditional hydrocarbon combustion process is to do away with the explosions and the moving parts. In the fuel cell approach, hydrogen atoms in the fuel enter from one side of the cell, oxygen molecules from the other side. They meet in the middle under the auspices of a catalyst, a matchmaking chemical that allows the direct formation of electricity (no turbines necessary), along with some water. Fuel cells have been around for a few decades for niche applications. They come in a variety of sizes and, since their pollutant products are minimal, they come in handy where emissions standards are a major consideration. Furthermore, advanced models are extremely energy efficient. Why aren't they more prevalent then? They produce low voltage, their power has to be converted from DC to AC, and the more efficient models tend to be expensive.

Can fossil-powered machines be made better still? Well, it would be nice not just to reduce but to eliminate pollution from the generation process. In one approach, coal would be turned into a synthetic gas in a way that allowed for the removal of troublesome chemicals, such as nitrogen and sulfur, even before any combustion takes place. In this scheme, going by the cumbersome name of integrated gasification combined cycle (IGCC), energy efficiency of 40 percent would be possible.

Gasification also makes it easier to remove carbon dioxide before it escapes into the air.[5] In so-called zero emissions power plants, the carbon dioxide would be pumped into a long-term storage zone, perhaps beneath the seafloor or in underground cavities that had lately been vacated by petroleum or natural gas. Undertaking carbon capture and storage (also called sequestration) would reduce the overall energy efficiency of the power plant, but the dire threat posed by smokestack emissions would be greatly reduced. The coal industry—or as Lewis Mumford liked to call it, "carboniferous capitalism"—could survive on a more sustainable footing.

Imagine this scenario: Carbon atoms (bound up in methane, say) come out of the ground in Texas, are piped to Illinois for combustion in some future zero emissions plant, and then, following combustion and locked into a carbon dioxide molecule, are returned by pipeline to Texas to be interred in the ground in a deep cavern. This would be a better place for that atom. It wouldn't be up in the sky, where it might trap a little too much sunlight and, through some weird greenhouse mechanism, force an unnaturally nasty turn in the weather.

CUTTING BUTTER WITH A CHAINSAW

A growing share of electricity is being made in medium- and small-sized generators, but a majority of power is still produced the old way—in large generators like Big Allis. Studies show that these larger machines are less dependable on average than smaller plants in terms of downtime.[6] Nevertheless, the leviathans are still a dependable presence on the organizational spreadsheet. A big plant with a big number of megawatts is easier to fathom for planning purposes than a bunch of smaller units, especially if, as in the case of wind turbines or solar cells, they're dispersed in scattered locations and subject to variable conditions such as wind speed or available sunlight. These views are changing as the smaller units' efficiency and reliability win them acceptance into the grid club.

For Amory Lovins the restructuring of the power industry over the past decade was something of a distraction. Yes, it changed how electricity is marketed, and how innovation is rewarded, and how energy companies relate to federal and state regulators, but it served to complicate an even more important movement under way: the prospective evolution, or devolution, back toward smaller *distributed* forms of power generation. Lovins draws attention to the huge mismatch in energy scale between the powerhouse level, generally 1,000 megawatts (a billion watts), and the home level of a kilowatt (a thousand watts). The main grid delivering electricity to your doorstep is like sipping water from a firehose, Lovins says. There is bound to be lots of wastage.[7] Larger plants no longer have the great advantage of efficiency (lower-priced electricity) but still retain many of the disadvantages of being large.

Here is a quick list of factors that go into the new small-versus-large math. Although construction costs per kilowatt of output may still be lower for the larger plants, smaller plants are, on average, available for service more often, have a shorter lead time, are easier to site, and are usually easier to finance.[8] Consequently, for the crucial task of planning future power demand, the plants can more easily be built as they are needed, "There is less time for reality to diverge from predictions," Lovins says. There is less room for regret. Errors in anticipating future loads are reduced.[9]

Two prospective changes are at work. One is a reversion to smaller plants. The other is a possible return to shorter transmission distances. This shrinkage of the grid would represent a great historical reversal. The ability to send alternating current electricity over great distances allowed power to be made in one place and used in another place. Niagara could dispatch power all the way to Buffalo and later to New York City. This expanded capability, in which large territories hundreds of miles across became unified energy *realms*, practically redefined gridness. This unity, this oneness, was manifested in some countries like Britain and France in the form of consolidated national grids. The dark side of consolidation is that, when part of the system goes down, a much larger region might collapse as well owing to the tightly coupled nature of the extended grid. Prime examples are the great Northeast blackouts of 1965 and 2003 and the collapse of the French grid in December 1978.

One problem with sending power over long distances is the loss of energy along the way, which can be as much as 10 percent. Another problem: transmission bottlenecks, which arise when too much power is being sent such long distances to wholesale customers on the far side of an increasingly congested network of transmission lines. Why not just build more lines? Because they're expensive to construct, politically painful to plan ("Not in my backyard"), tricky to finance (in a volatile business with huge price swings), and difficult to regulate (conflicting state and federal statues).

Taking into account all these various costs—transmitting, routing, transforming, movement across jurisdictions, the accounting that goes with various levels of buying and selling—we arrive at a curious development. It might now cost more to deliver electricity than to make it.[10] This is true for many things in the top-heavy merchandising world where packing and sending and selling become the tail that wags the dog.

What can be done? If a family were contemplating taking a driving vacation on the interstate highway system, with the prospect of traffic jams, lots of tolls to pay along the way, lines at the gas pump, and the engine making weird noises, it might be tempting to stay home. This doesn't sound very glamorous. Where's the adventure?

But electricity doesn't need to be glamorous. Power is power. What if the grid stayed home? This is the microgrid concept, a scheme in which a confederation of small electricity generators and electricity users are wired up to a semiautonomous network. The users could be homes, factories, or farms. The generators might be wind turbines or housetop solar cells, fuel cells, or basement gas-powered turbines with ratings as low as a kilowatt. A microturbine of that size might not be as energy efficient as larger models, but if the heat in some of these small machines could be captured and exploited for warming a house, say, or to contribute to a factory process, the economy of this microgrid might start to rival that of the macrogrid. The goal would be to attain all the good features of *big* while retaining all the virtues of *small.* The microgrid would be to the grid what desktop publishing is to mainline publishing. The hardware and software are now becoming available for creating a desktop grid.

The goal of the participants in a microgrid is not necessarily to secede from the macrogrid. The dispersed grid would still continue to be connected to the central grid through a common hookup. The electricity generated internally would be used internally. But outside electricity could be imported as needed. If, for example, the microgrid's solar cells weakened when clouds passed in front of the sun or its wind turbines became limp when the wind died, or if the price of gas use in the microturbines were to shoot up, then extra power could be brought in. The microgrid would appear to the macrogrid like any other customer.[11] There might even be times when the microgrid had an energy surplus—more than it could use by itself—in which case the current would actually flow back into the commercial grid. The utilities themselves are more open nowadays to alternative power concepts; *small* and *renewable* are not necessarily dirty words anymore.

THE HOME GRID

A single home can be a microgrid unto itself. To see how this works, we will now inspect a model home in the town of Takoma Park, Maryland, dedicated to the idea of clean or green energy use. In hifalutin terms green means disturbing the universe as little as possible on your journey through life. How much can an individual disturb things? Well, the doings of each citizen in one country (the United States) are attended by the release each year of about 12,000 pounds of carbon dioxide, which (after trees and oceans absorb a certain amount) can be considered uncollected garbage sent into the air. The atmosphere is an arena for complexity-driven chemistry, and extra CO_2 molecules are just so many sand grains that might trigger future avalanches in weather. Green discourse can sound tiresomely moralistic, but considering its impact on world affairs, the making and using of electricity—and energy in general—have necessarily become moral issues.

The clean-energy home isn't just a stunt or an experiment but a place where a family lives. The first thing you see, coming out of the cold blustery morning into the warm living room is a curious enclosed fireplace fed from above by a hopper filled with corn kernels. Now, burning corn to make heat may well release as much carbon dioxide as burning an equivalent amount of coal so, right off, how is this an example of "clean energy"?

It's because the corn, as it was growing, was *absorbing* carbon dioxide from the air, as all green plants do as part of their regular metabolism. In an important sense the corn has, in effect, prepaid for the carbon pollution it makes at the combustion end by reducing a like amount at the growing end. That's why certain municipal biomass power plants—burning things like pecan shells, apple pomace, or straw to make electricity—can be considered a renewable source of energy. This is the reasoning behind ethanol, auto fuel made from grain or sugar.

Once you produce all this warmth, you need to hoard it. You can do this through the use of thick insulation in the walls and ceiling and high-tech windows that let light in but don't let heat out. Another crucial step in using energy sparingly is to have the right appliances. So, for example, the clean-energy home uses high-efficiency fluorescent bulbs (using one-fourth the electricity for the same amount of light) and the kind of refrigerator requiring half the power of models only 10 years old.

The establishment of a lean-consumption house is not an act of defiance. It's like making the decision to eat sensibly. You increase your chances of feeling better and living longer. Putting your home on an energy diet is one way of shaping the material circumstances of your life and of society itself. Instead of merely taking energy out of the socket as it comes, you can do something about it. Does this mean you have to make your own energy? Isn't that what they did in the old days? Time-consuming scrounging for wood and making inefficient fire in the home—isn't that what mass-distributed electricity and gas put to an end? Doesn't the modern network work pretty well?

Yes, the utility system does work pretty well, and it can be made even better, say proponents of renewable energy sources. Forget for the moment the argument over whether these continually replenished sources of energy can ever be incorporated on a large scale into standard grid operations. They can at least be employed, where appropriate, on the homefront. At the Takoma Park home, for instance, the water heater still is warmed with gas energy. But the water first is diverted to the roof, where it slides through pipes warmed by sunlight. The conventional water heater is not eliminated but rather *assisted* by the solar heating device which, in this case, contributes about 60 percent of the energy needed to make the water hot.

If we think of the model home as a showcase for green energy, the main display would be the photovoltaic panels parked on the roof not far from the warm-water pipes. These solar cells imitate photosynthesis, the marvelous process that serves as no less than the basis for the entire food chain on earth. Photosynthesis happens when a ray of light from the sun strikes a chlorophyll molecule in a leaf, liberating an electron at one end of the molecule. The electron gets handed about like a dish going around a dinner table. Through a series of biochemical reactions, complex carbohydrate foodstuffs are built up. Thus the plant does not contain the energy: It *is* the energy. An animal eating the plant will inherit this energy-packed material, which will in turn be used to build up the animal's own body and so on up through the food chain. We are all rearranged atoms and repackaged sunlight.

A solar cell, also called photo cell or photovoltaic cell, is an artificial leaf. It tries to reproduce nature, which is often a good strategy in engineering since, after all, natural systems have had millions or billions of years to adapt and improve themselves. In a solar cell the absorption of light occurs not in a chlorophyll molecule but in a thin layer of semiconducting crystal. As in a green plant, the sunlight liberates an electron. Instead of helping make glucose, the electron in a solar cell moves into an external wire where it can be used to charge a battery or power an appliance. The solar electricity runs to a box in the basement where it is converted from direct current into alternating current and sent pulsing at the same 60-cycles-per-second throb as the regular gird.

The advantages of getting electricity this way are that there's no pollution, and the fuel (sunlight) is free. The disadvantage? The solar cell is expensive to buy. So it was with most early electrical apparatus. The economics of solar cells are improving and unit costs are coming down, but the price per kilowatt-hour is still too high for solar to play an important role right now. In the case of the clean-energy house, tax credits helped make the solar cells affordable. And as soon as they're operating, they start paying for themselves by reducing the monthly utility bill. It might take a number of years, but eventually the rooftop grid will have paid for itself.

The cells work well, so well that on some sunny days the power coming from the roof more than fulfills the household energy budget and electricity can actually be sent out into the macrogrid. (Some but not all home power hookups allow this "net metering.") At that moment you really have a new thing in the electricity business: homes not just taking power *from* but also giving power *to* the grid. The river of current has reversed, and the wallside meter runs backward. The homeowner gets a breath of satisfaction that in a small way she gets to influence the way the system works. Every home is a castle, and now every home can be a minor power station.

The combined effects of restructuring and engineering innovations have changed the power industry. The vast preponderance of central generation in massive, largely fossil-fired or nuclear plants has been reduced. For instance, in the year 2005, decentralized sources—including small- and medium-sized gas turbines used for cogeneration of heat and electricity, plus renewable generation via wind, photovoltaic, biomass burning, and small hydroelectric facilities—delivered more power worldwide than nuclear reactors.[12]

The owners of the clean-energy home don't live a monastic life. Even with the rooftop solar panels, they have need of the regular grid. Those who seek to use low-impact energy still generally resort to gasoline-fueled cars. They fly on airplanes sending fumes into the air. They still require macrogrid electricity made in fossil-fired megaturbines. What can they do about this? What they do is purchase wind-powered electricity. No, they don't have windmills on the roof. And they are not connected directly by wire to the nearest sizable wind farm 140 miles away on a West Virginia mountain ridge, but they *are* connected in a way.

Wind power, like photovoltaic power, is clean but still more expensive than that produced in old-fashioned steam-electric plants. On a volunteer basis, however, the customer inclined toward green energy can pay a little more on the monthly bill to offset the higher cost of wind power. Why willingly pay more? The symbolism is this. Even as the customer continues to use the conventional grid, he or she is at least reaching out to, or helping sustain in some measure, a cleaner approach to power production. Paying extra money for what is, physically, the same generic electricity that everyone else gets, involves an act of imagination. You envision and partly promote a departure from business as usual. By paying for the cleaner production, you're keeping thousands of pounds of pollutants out of the air. You're saving a tree somewhere from succumbing to death by acid rain. Smog, and the wheezing that comes with it, will be reduced by an increment. This is the green view of electricity.

One of the fastest-growing renewable resources is wind power. In the United States acceptance of wind power has been slower than in Europe but is now catching on quickly. The not-in-my-backyard problem is serious; people who own land with a nice view don't want that view interrupted by gangly propellers whirring away in the distance. In places like Long Island and Massachusetts, where many people (including influential politicians) might otherwise be friendly to green energy, opposition to coastal wind farms (which compromise oceanfront real estate) has been fierce. The solution might be to build the wind turbines farther out to sea, where they are beyond sight or at least only a dim blot on the horizon. The cost of wind power is coming down. In some breezy places like Denmark wind power already accounts for as much as a fifth of electrical production, and that fraction is expected to grow. Some countries, especially in Europe, are counting on wind energy to help them meet their pledge, under the Kyoto Treaty, to reduce the amount of electricity made by emitting carbon dioxide.

On the way out of the clean-energy home, you see two cars parked in front, each meant to illustrate the energy reform idea as it applies to transportation. One car is entirely electric. An electric motor converts more of the input energy into motion than does an internal combustion engine, and yet you see few electric cars. Except for dedicated fleets of vehicles, such as the postal service would use, electric vehicles don't yet have a nationwide or worldwide network of service stations for recharging. Also, the batteries needed by these cars are expensive and difficult to dispose of.

The other car out front of the home, a so-called hybrid model, has two methods of motion, one electric and one combustion. This clever dual design confronts the infrastructure problem head on. The gas-fueled engine, as it moves you down the road, recharges a battery feeding an electric motor, which can take over some of the propulsion chores in order to exploit its higher efficiency. Gas mileage goes up; pollution comes down. The hybrid approach used to be a novelty but might soon become a mainline (maybe *the* main) automotive architecture.

The car you don't see out front, because it is extremely rare and not out of the experimental stage, is the hydrogen car. Some things to do with hydrogen are quite simple. It is the lightest and simplest of all atoms. It's the first entry in the periodic table of elements and is the most common commodity in the universe. Combined inside an engine with oxygen, it is the cleanest of fuels, producing only water vapor as an exhaust. It's that simple: The hydrogen-consuming engine—a variation on the fuel cell design—produces electricity and water. The electricity could be used to power a car or even, at a moment of peak need, be plugged into a socket and sent into the general power grid at rates as high as 20 to 40 kilowatts. In some future society the aggregation of hydrogen cars on the streets, thousands or millions of them, could conceivably constitute an immense auxiliary ambulatory network of electric generation, a sort of rolling grid.[13]

Everything else to do with hydrogen is complicated, at least for the present state of technology. Hydrogen is relatively difficult to store economically, and there is currently no means to distribute it on a wide basis. The main complication is procuring the hydrogen in the first place. That's because hydrogen atoms are often in the company of other atoms. The most economical method right now is to extract hydrogen from natural gas.

It bears repeating: Electrons are present in wires to begin with. They are *carriers* of electrical energy. A generator doesn't create the electrons; it *pumps* them along. The same would be true of hydrogen. Hydrogen atoms were made shortly after the Big Bang billions of years ago. They're already around. If extracted and pumped they too, like electricity, would be a carrier of energy. Think of energy flow via hydrogen as "hydricity."[14] You take hydrogen, add oxygen, and *voila!*, you have electricity. It's like having freeze-dried electricity. It sounds great, but the technology to make it all work isn't available yet. Many experts believe the use of hydrogen as a fuel or as a serious electricity substitute is decades away. Recall my earlier idea for an "Ohio Valley Authority," an imaginary hydrogen-power equivalent of the TVA. Maybe we need an effort on that scale.

In the meantime, how do we energize the electrical grid? That's the central issue of this chapter. Gas turbines have come to shoulder a greater share of U.S. power production and in other countries too, but gas prices have been volatile. Energizing the grid with renewable energy looks more hopeful than it did. Prices are coming down, and options are multiplying. Wind power especially is coming on strong. But the absolute amounts—a few percentage points of the U.S. power market—are still scant.

Will this cultivation of renewable sources and an ever greater crusade to eliminate energy waste be enough to meet future power needs and displace coal from its majority role in power production in the United States, China, and many other countries? Amory Lovins and other champions of renewable energy and efficient usage believe the answer is yes. Others aren't so sure. For them the numbers don't add up. Consequently, there has been a renewed discussion in industrial circles and in government of an old-fashioned approach to large-scale power production. Old but free of airborne pollutants. The technology is well known and the fuel relatively cheap.

It is time to talk about nuclear power.

CRACKING ATOMS

For some the nuclear option is the Rock of Gibraltar, a reliable way of extracting a lot of carbon-dioxide-free electricity from a tiny amount of enriched fuel. It is the once and future stalwart producer of baseline electricity. For some grids even now, nuclear reactors do the heavy grunt work of sustaining the grid around the clock. Many of the largest generators in the world, those that produce the most kilowatt-hours, are pressurized-water reactors.

For others, nuclear operations are a darkness, a dread prospect to be rejuvenated only if all other alternatives prove inadequate, and maybe not even then. However unreasonable it is to compare nuclear weapons with nuclear power plants (a reactor cannot explode like a bomb), visions of mushroom clouds seem to loom in many minds at the mere mention of the word *nuclear*. To them, greenhouse effects and clouds of smog set loose by burning coal seem less dire by comparison.

Should the grid be energized by nuclear reactions? To have a closer look at this tangled issue, we undertake another inspection. Visiting a nuclear reactor is not like going to the castle of the evil witch in the *Wizard of Oz*. Here at the Indian Point facility in New York State, there are no crenelated battlements, no flying monkeys, no dirgeful singing. There are, however, plenty of razor-wire fences and guards with automatic weapons. This might be the most heavily defended industrial facility in the United States.[15]

Nuclear activity at the site dates back to 1962, when Consolidated Edison started operating its first reactor. This machine didn't last long because new safety guidelines required prohibitively expensive retrofits, but two other reactors, Units 2 and 3, came into operation only a few years later, one sitting on either side of the retired Unit 1.

It is to this nuclear park, some 40 miles north of New York City on the east bank of the Hudson River, that we have come for a look. To reach the operational center you must negotiate a series of barriers: checkpoints, offices for visitor badges, fences, concrete traffic baffles, body-armored officers, metal detectors, trunk search, and the scrutiny of men poised in the kind of observation towers one sees at the perimeters of penitentiaries. Out on the river a Coast Guard ship patrols. Once inside it is all smiles, informative discussion, helpful pamphlets, and a guided tour around one of the premier nuclear operations in the country, where 2,000 megawatts of electricity is produced.

Indian Point is no longer owned by Consolidated Edison of New York but by the Entergy Corporation of New Orleans. Restructuring strikes again. One of those huge energy companies that have come to dwarf many of the old mainline utilities, Entergy controls 30,000 megawatts of electricity, some of which it distributes to its own customers, some of which it sells to other distributors like Con Ed under long-term contracts, and some of which it sells on the spot market for as much as it can get.

While many power companies were eager to get out of the nuclear business, Entergy embraced it in a big way. It owns a fleet of 10 reactors in six states and holds down costs by sharing parts and experience among the sites. Although several decades old now, nuclear technology is still a youthful field in comparison to coal and gas combustion technology. Actually, nuclear and conventional fossil power plants have a lot in common. Both use a lot of 19th-century physics. Both employ high pressure fluid-dynamic practices; first make a lot of steam and then use it to keep a turbine spinning for days, weeks, or months at a time. Both use the same electrodynamic practices; the spinning shaft sets up winds of electric force that are sent through wires to the far-flung corners of the grid. Where nuclear and conventional generators differ greatly is in their manner of making steam.

Before going any further at Indian Point, you must visit the health physics department, where you receive two separate devices to be worn around or near the neck. One measures the total amount of radiation you might encounter (although little or none is expected) and is read at the end of the day. The other fancier unit measures radiation moment by moment and radios the results back to this office. It is in the realm of health physics where 19th-century science gives way to 20th-century science.

In a fossil fuel furnace—the 19th-century kind—hydrocarbon molecules (in the form of coal or natural gas) are mixed with oxygen molecules in that age-old combustion reaction that results in fire. In the fission fuel furnace—the 20th-century kind—uranium atoms, struck by subatomic particles called neutrons, fly apart in a reaction first observed by humans only a few years before the start of World War II.

In the fire reaction, no atoms are destroyed, only rearranged. In the fission reaction, atoms are emphatically dismembered. That's what fission means: to break apart. The leftovers from this microscopic devastation consist of a couple of daughter atoms, some surplus energy, and a few extra neutrons that can proceed to trigger further fission reactions. The fire reaction involves only the outermost part of the atoms, the part where electrons reside. The fission reaction all has to do with the nucleus, the innermost part of each atom. If an atom were the size of Madison Square Garden, its nucleus would be the size of a sugar cube. Nevertheless, far more energy is latent within that tiny nucleus—the sugar cube—than in the whole rest of the atom. A carbon atom has a nucleus too, but it is not ripe. It can't fly apart the way a uranium nucleus can—hence the great disparity between carbon fuel and uranium fuel. The fission reaction releases, on average, millions of times more energy than the fire reaction.

Let's differentiate between two kinds of nuclear reactions: radioactivity and fission. In radioactivity the nucleus casts off a tiny fragment of itself. In fission the sundering is more complete—the nucleus essentially breaks in half. Radioactivity, insofar as it produces penetrating particles, can do biological damage and is potentially dangerous. Fission packs much more of a wallop and has generally been exploited for only two specialized applications: nuclear bombs and nuclear reactors.

In a fire-filled furnace, the kind used by Edison at Pearl Street or by Big Allis at Ravenswood, most of the atoms in the fuel, once combusted, depart the premises. Except for some ash, the spent atoms go up a stack and drift out to sea, or over a forest, or some distant kingdom or republic on another continent. This long-distance dispersal of molecular pollutants is a matter of great importance and diplomatic discussion since the molecules might be killing trees or melting the polar ice caps.

In the fission-filled furnace, the kind used at Indian Point, the byproducts are very different, so different that there needs to be a health physics office at every reactor. After making heat, the atoms in the fissionable fuel don't go anywhere. Some of them have been transformed into other atoms (uranium atoms break into such secondary atoms as iodine or strontium), but these daughter products do not fly up any stack. They stay embedded in the fuel material. And this fact is both welcome and troubling: welcome because there is no greenhouse gas sent up to be stored in the atmosphere, troubling because the waste that stays behind, those radioactive daughter atoms that never left home, must be stored for years in special underground, water-filled tanks nearly as heavily monitored and protected as the reactor core itself or in "dry cask" containers.

On to the power-producing part of the plant. The turbine room looks the same as the one at Big Allis. Loud almost to the point of deafening (you wear earplugs at this point) and hot (steam races through pipes beneath the floor), this is the place where thermal energy is converted to electrical energy. This is straight 19th-century physics. Much of the power made in this room ends up in the same form as Big Allis's power, 345-kilovolt electricity, and goes to the same New York customers. You can't tell one volt from another. When you turn on your computer, nuclear-made and fossil-made electricity mingle indistinguishably.

Nuclear reactors account for about a fifth of the electricity made in the United States right now; the worldwide number is similar. At one time many thought the fraction would be much higher. Others, citing notable nuclear accidents like Three Mile Island and Chernobyl, expected the fraction to be much lower by now. Even though the financial burden of building or starting (and later abandoning) reactors back in the 1970s and 1980s was so great (and still influences investment decisions decades later), and even though no license has been granted for new U.S. reactor construction in decades, the nuclear option is still a considerable factor. In France it constitutes 80 percent of the power industry.

Today Unit #3 soldiers along at near full capacity. Its turbine room is full of noise and vibration but practically devoid of human traffic. Over at Unit 2, turned off for refueling, it's the other way around; its turbine room is quiet but filled with activity. For the 28 days of downtime, Unit 2 will receive $100 million worth of improvements, not only replacement fuel but also a new high-pressure turbine and other equipment that will add 40 megawatts to the power output. If we're lucky, we'll be allowed into the reactor room.

The main task at hand is to install new fuel in the core. What does uranium fuel look like? It's metallic in form and shaped into half-inch-wide pellets, which are stacked like poker chips inside rods made of zirconium, a metal with a high melting temperature. The rods in turn are gathered into bunches called assemblies. Finally, the assemblies are loaded into the reactor core vessel, a stout pressure cooker inside of which fission heat warms water circulating through the fuel assemblies.

Uranium isn't just uranium. Like Toyotas, it comes in several models. The main version is U-238, meaning that each uranium atom has a nucleus containing a number of neutrons and protons adding up to 238. A slimmer version of uranium is U-235, whose nucleus contains several fewer neutrons than U-238. Both cousins have the same chemistry but very different nuclear behavior: U-238 is radioactive, meaning that over millions of years there is a good chance it will lose a small portion of its mass in the form of radiation. U-235 is radioactive too. However (and this is the big difference), it is also fissile, meaning that if it's struck by a neutron of just the right energy the nucleus will break in half like a pinata, not millions of years from now but immediately.

This property of U-235 makes it valuable as reactor fuel. It is the U-235 nuclei that provide the potent energy to keep a reactor going. The reactor could go for a year or more without stopping, and this is why nuclear-powered vessels can stay at sea for long periods. Energy deposited in the U-235 nuclei billions of years before in a supernova explosion of a star in deep space is, in the reactor, reincarnated as tidy electric waves sent south to Brooklyn.

After a while the U-235 atoms with all this pent-up energy undergo fission and the fuel rods lose their effectiveness. How are the fuel rods like members of the U.S. Senate? Because roughly every two years one-third of them come up for replacement. Every two years about one-third of the fuel assemblies in a reactor are replaced with fresh rods in which the U-235 concentration is at the 4 percent level rather than the 1 percent level in naturally found uranium samples.

The control room (to be more exact, the full-scale replica of the control room, used to train new operators) looks like the one at Big Allis. It has the appearance of an air traffic control center, and indeed the man in charge refers to the bank of dials in front of him as the "flight panel." Just as many commercial airline pilots get their start in the Air Force, many operators of commercial reactors get their experience in the Navy, where they pilot the nuclear reactions that propel submarines around the world.

Indeed, the nuclear reactor business floats in water. Water is what keeps the fuel in the core from getting too hot. Water is what slows neutrons to just the right velocity for keeping the fission process going.

And water is what carries heat away from the core, typically to make steam in a secondary set of water-filled pipes. The rate at which the fission takes place can be moderated by spiking the water with boron atoms, which absorb neutrons, or by inserting control rods that also soak up the neutrons. Put the rods in and the fission reactions stop. The nuclear candle is snuffed out. Pull the rods out and the reactions can resume. The rods are designed in such a way that no matter what kind of accident might occur, even a loss of electrical power at the reactor itself, the rods will automatically insert themselves into the core and shut off the flow of neutrons. Starved of neutrons, the nuclear fission process stops within a few seconds.

In the United States right now 104 reactors are on duty. About 10 percent of these are owned by Entergy, making it one of the largest nuclear operators around. And Entergy's corporate reports indicate that it is making a nice business out of its reactor fleet.

A power plant sells more than power. Consider this scenario of a utility making an arrangement with an independent generator: "We want 200 megawatts. Not right now. Maybe not at all. But if we want it, we'll need it in a hurry." What the utility is buying is not power but the capacity to get power. It's buying backup power, contingency power. With this contract in hand, the utility might not have to have one of its own generators standing by. This contingency, standby power, is one of the services provided by Entergy.

We're in luck. Sticking one's head inside the containment building, but not much more, will be allowed today. This inner sanctum protecting the reactor is the last of the day's barriers to get past: the 4-foot-thick wall surrounding the reactor pressure vessel that holds the fuel assemblies. Standing on the threshold is as far as we'll get. Any farther and you'd have to don the protective suit worn by the workers scurrying about inside. In the distant part of this cavernous room one can spot the removed top of the reactor. Again, the large room swallows the illumination; lamps are on everywhere and yet it all seems dim. Machines cast dramatic shadows, as if this were some Halloween party. From the doorway it is impossible to see the fuel assemblies being put back into the reactor, but this meticulous work can be watched on numerous video screens.

Questions of safety and security are ever present. For instance, an engineer, seeing us look over at the reactor top, explains that the caps on the pressure vessel here at Indian Point are nothing like the one at the Davis-Besse power plant in Ohio. There a cap was almost eaten through with corrosion. Had the crack broken all the way through the vessel wall during normal operation, the reactor's cooling water, laden with radioactivity, might have spewed out into the containment room. Bad as this would have been, the mess at least would have been contained within the reactor room. That's why it's called a containment structure. But the crack was spotted in time.[16] Besides, that was Ohio and this is New York.

The Indian Point containment structure can sustain the direct hit of a commercial jetliner without rupturing, one is solemnly told by plant personnel, although no direct test of this proposition has been made. Standing in the doorway, the walls look sturdy. Still, it's creepy to think that only a few days after Entergy purchased this reactor in September 2001, a hijacked jetliner passed overhead, or not far off, on its way down the Hudson Valley toward a rendezvous with the World Trade Center.

NUCLEAR REACTIONS

Physicist Richard Garwin, credited with being one of the developers of America's hydrogen bomb in the early 1950s, distills the essence of nuclear power down to a few absorbing facts. To operate a typical 1,000-megawatt reactor for three years, he says, you would install about 75 tons of fuel. At the end of three years, having supplied about 25 billion kilowatt-hours of energy to upward of a million people, you would remove 75 tons of spent fuel, minus 3 kilograms, or about 8 pounds. Those 3 kilograms of matter—originally solid material—were turned into pure heat energy, not through any chemical burning process but by nuclear reactions. And of that only about one-third ends up as usable electricity. In other words, three years' worth of electricity for a good-sized city came from the conversion of a kilogram (between 2 and 3 pounds) of uranium-235.[17] That, in a nuclear nutshell, is why reactors exist.

As you can see, nuclear power is a fact-rich subject. It's also opinion rich. Take, for example, the matchup of fossil fuel and fissile fuel power generation. Fossil generation, through its worldwide respiratory effects and coal-mining accidents, in an ordinary year, kills more people than the accumulated radiological effects of all past fissile generation. Yet the very idea of a nuclear disaster, with images of the Chernobyl explosion spreading contamination over a wide area and the vague blurring of nuclear power with nuclear weapons in many minds, seems to inspire a visceral dread disproportional to the real-life mortality rates. The design of nuclear reactors and the comparative risk analysis of fissile and fossil power production is so hedged with actuarial factors, game theory mathematics, and probability science as to make the subject practically a branch of quantum physics, which, technically, it is.

Here we arrive at the most vexed question concerning the production of electricity. Whether to launch a major new round of advanced-design nuclear reactor construction in developed countries is a many-billion-dollar decision. Because of its implications for the environment and for national security, this is also a foreign policy issue of prime importance. To ponder so contentious and weighty a subject, I plan to examine a variety of perspectives.

The first view belongs to the owner of the Indian Point reactors, Entergy Corporation. Its motto is: "Safe. Secure. Vital"—safe, as in maintaining a good safety record and in keeping its inventory of radioactive material where it should be; secure, as in having a multilayered defense system against theft, intrusion, or mayhem; and vital, as in accounting for a sizable fraction (up to one-fourth on any one day) of New York City's power needs.[15]

The second view, that of Riverkeeper, a private conservation organization seeking to close Indian Point, is 180 degrees apart from that of Entergy. Indian Point and its radioactive store of isotopes is, to Riverkeeper, another Chernobyl waiting to happen. Nuclear proponents never tire of pointing out that the core meltdown and explosion at the Chernobyl site in Ukraine, history's most serious reactor accident, was more lethal partly because it lacked a sound containment structure, resulting in a wider dispersal of radioactive material outside the immediate compound. By contrast, reactors in most other countries always have a containment roof over their heads. Riverkeeper, without saying how Indian Point's fissile material would be released into the environment (by accident or terrorist attack), does paint a vividly apocalyptic scenario, one in which 40,000 short-term deaths and as many as a half-million long-term deaths would accrue.[18]

A third view summarized here is that expressed by Richard Garwin and Georges Charpak in their book, *Megawatts and Megatons: The Future of Nuclear Power and Nuclear Weapons.* Charpak won a Nobel Prize for his development of instruments to detect subatomic particles. Garwin is a long-time Pentagon consultant and critic, expert on numerous defense matters, a former research fellow at IBM, and a recognized authority on condensed-matter physics. Garwin and Charpak, who know more than most the perils, technology, and costs associated with nuclear power, believe that fissile generation of electricity, even with all the complications, can have a viable future but only if certain things happen. The first requirement is a safer reactor design. What about current safety? Are reactors safe right now? "Risks of normal operation are at present acceptable," Garwin says.[19] Fuel availability, he believes, seems secure. As for spent fuel, it can pile up at current temporary storage areas (often right next to the reactors) for decades to come. Another significant development over the past dozen years is the growing reliability of reactors. They are up and running a greater portion of the time, and the rate of unplanned shutdowns at U.S. reactors has been declining.[20]

Arguments in favor of continued or expanded use of reactors often return to a comparison of the relative risks of power production. The argument goes this way: Society tolerates many risky things. Here are a few. Putting gaseous pollutants into the air (from burning coal, say) increases respiratory disease and might be triggering deleterious climate change. Eating fatty foods, with the associated cardiovascular peril, is a risky habit. Driving an automobile is hazardous; more than 40,000 highway fatalities annually in the United States, however, haven't dampened the desire to drive. Seen in this light, the cracking of atoms in order to obtain electricity seems tenable to Garwin. His considered opinion comes down to this: "Normal operation and even occasional disaster still leave nuclear power with a health benefit over competing sources such as coal."[21]

Another important and related issue (some would call it the highest international security issue of our time) is nuclear proliferation, the spreading acquisition of nuclear technology know-how and of nuclear materials, whether by nations or rogue organizations. The U-235 used in commercial reactors and one of the byproducts of reactor operation (the plutonium atoms created when neutrons strike U-238 nuclei) make spent fuel rods a natural repository for exactly those materials needed for assembling potential homemade nuclear explosives. Consolidation of vulnerable nuclear inventories and greater international monitoring of the flow and handling of radioactive and fissile substances are urgently needed.

All these points were made in a recent study prepared by scientists and engineers at the Massachusetts Institute of Technology. Their report described efforts to reduce by 10-fold the risk of a reactor accident by making reactor design simpler and foolproof.[21] The MIT report and the nuclear power industry itself readily concede that for the amount of electricity you would get out of it, building a new reactor is too expensive when compared to power from already-operating coal- or gas-fired plants. The cost differential narrows when the price of a new nuclear reactor plant is compared with the price for a new coal- or gas-fired plant. The disparities can substantially be leveled out further by assessing a carbon tax on polluting fossil-fuel-produced electricity.[22]

The financial, regulatory, and public relations obstacles for a nuclear renaissance in the United States are formidable. But so is the need to reduce the emission of greenhouse gases during the production of electricity. Consequently, various power companies are now seriously pondering whether to seek a license for building new reactors.[23]

We can't satisfactorily answer the question, Is nuclear power safe? (compared to what?), so we have to keep searching for additional ways of viewing nuclear reactors in the context of other complicated machines. Charles Perrow's book *Normal Accidents* looks at how complexity—that behavioral syndrome featuring tightly coupled and rapidly interacting parts—plays out in specific systems. He sorts many specimen examples (post offices, hydroelectric plants, coal mines, etc.) on an organizational chart whose horizontal axis measures how the system components interact, from linear (the output being proportional to the input) up to unpredictably complex. The vertical axis represents the amount of component coupling, from loose to tight. If you will recall from Chapter 6, the vulnerability of complex systems to avalanche failures is related to how quickly and how strongly the system components interact.

And the system in the upper-right corner of Perrow's chart? The system with the extremity of tightly coupledness and nonlinear interactions? The system with the greatest propensity for complicated, unpredictable behavior? On Perrow's chart it is nuclear power plants.[24]

This does not mean that nuclear plants are doomed to having serious accidents, only that the linkages and pathways of causation are plentiful. Furthermore, we could point out that other complex systems are vulnerable to accidents too. Petrochemical plants are highly coupled and nonlinear systems, but they have been around a lot longer than nuclear plants and there has been more time to gather precautionary experience.

Despite such chemical industry disasters as the one in Bhopal, India, in 1984, when the escape of a dangerous gas from an insecticide factory led to more than 2,000 fatalities, the scope of hypothetical nuclear accidents would seem to be higher than for chemical accidents. Consequently, Perrow says, we have a Nuclear Regulatory Commission for spying on reactors but no such thing as a Chemical Regulatory Commission for snooping on refineries.[25]

THREE CONTENDERS

We've made three site inspections in this chapter to illustrate three approaches to energizing the grid. Big Allis epitomizes the fossil method for electricity production. Since a lot of carbon atoms are burned this way, let's designate this path with the letter C. Then there was the clean-energy house, exemplifying the renewable approach (denoted by the letter R), which includes things like wind and solar power and energy gained (or energy use avoided) by greater efficiency efforts. Finally, there is the nuclear (N) approach used at Indian Point.

In Amory Lovins's designations of hard and soft approaches to energy use, R represents the preferred path to energizing the grid while trying to minimize damage to the environment. Modular and compact, solar cells and wind turbines are tailor made for decentralized, distributed grids. N is the opposite. It is very hard. Nuclear reactors come in only one size: big. C is generally also big. Big Allis is big. But owing to all those polymorphous gas turbines—machines that can render kilowatts or hundreds of megawatts—C could go either way, allying itself with N in sustaining the traditional macrogrid scenario or with R in a hybrid macro/microgrid system.

There are, of course, a few generators that don't fit these categories. Not all renewable energy plants are small—biomass generators can be big. And not all small generators run on renewable energy—microturbines run on gas. For right now, let's concentrate on C, R, and N as if they were the only choices.

N produces radioactive waste but no gaseous pollutants. With C it's the other way around. R produces little waste. N's metallic uranium fuel was made in exploding stars billions of years ago. C's hydrocarbon fuel was made in the form of green plants millions of years ago. Meanwhile, R's fuel is sunlight pouring down now or wind set in motion by the warmth of sunlight six months ago. N's fuel can be made into a bomb (albeit with much extra costly processing), which makes it an inviting terrorist target. C's fuel cannot explode, but it can and often does enter into international security discussions. R's energy generally falls freely from the sky.

One of the major reservations to C—its prodigious release of problematic molecules into the open air—would be mitigated if gasification of coal could be combined with the capture and storage of carbon dioxide. The term "clean coal" would be much more merited in that case. Interring carbon dioxide has its own cost, but when you take into account that generation is only part of the overall expense, the actual rise in an average electric bill might be only about 20 percent.[26] Carbon storage technology is not yet proven on a large scale, however, and power companies seem reluctant to abandon old practices in the absence of an outright tax on carbon emission.

At this time federal legislators continue to postpone imposing a carbon tax, so the states are taking matters into their own hands. California recently decided to place ceilings on power-plant emissions.

This will be another of those great experiments in power production. If generators there can meet the new standards without substantial rate increases or loss of productivity then other interested parties (states, utilities, regulators, manufacturers, Congress) will take notice. Like American auto manufacturers in the 1970s, who at first railed against higher milage standards (standards that saved consumers billions of dollars in fuel costs), electric companies will be able to say, "That wasn't as hard as we thought."

Meanwhile glaciers are melting and reefs are dying. A large fraction of research papers on these subjects in refereed scientific journals say that the observed modifications are speeded by technological CO_2 greenhouse effects. And whose fault is this? Yours. The power companies are only doing your bidding in cranking out electricity by burning coal.

The studies mentioned in the previous section say that R should be encouraged in order to offset the polluting effects of C and because it's the right thing to do. They also suggest, however, that the anticipated rise in electrical demand in coming decades will necessitate renewed growth of N. Still another report, issued in 2005 by a U.S. Department of Energy advisory panel, makes essentially the same points. Why should we be more confident of nuclear power now than 20 years ago? More operating experience, a low reactor accident rate, a growing capacity factor (the fraction of time a reactor is up and running), typically 90 percent or better, and the prospective advent of an inherently safer and simpler reactor design are given as evidence.[27]

Recognizing the huge up-front costs of constructing a first-of-a-kind new reactor, and the great trepidations for a company or consortium being the first to seek a nuclear license in a third of a century, the DOE report suggests that the federal government pay up to half the cost of newcomer plants. One of the report's panelists, physicist Burton Richter, who won a Nobel Prize for his discovery of a subatomic particle, said that once you get the startup costs out of the way, and if you impose taxes on carbon emissions from fossil-fired plants, the nuclear approach becomes surprisingly competitive with the coal approach to energizing the grid.[28]

The marketplace should settle the issue. Amory Lovins believes there is sufficient renewable energy available (particularly wind power) to energize our burgeoning cities.[29] Furthermore, he says, although there might be sound security and environmental reasons for not using the nuclear option, the main drawback is that the construction and maintenance of nuclear plants will continue to be too costly, regardless of what Congress or the president might say in their public statements. The proof of nuclear inadequacy lies in the fact that private companies are not ready to invest their own money alone in the venture. When utilities start betting their own money alone, you might expect there is merit to nuclearism. "I'd watch Wall Street, not Pennsylvania Avenue," says Lovins.

Many subscribe to the view that government shouldn't pick winners and losers among technological alternatives, but in practice politicians can't help playing favorites. The tax code has always resembled the children's board game "Chutes and Ladders," in that it allows some players to glide easily toward their goals while others struggle against barriers. Skewing the exercise of capitalism can come in the form of overt subsidies such as R&D funding, tax rebates on products sold, guaranteed loans, and accelerated depletion allowances. Or they can be packaged more subtly as tariffs, lowered emissions standards, federal health benefits for employees, or ceilings on liability suits. Federal largesse, and the nation's energy policy, ought to be solemnly determined on the basis of merit but often depends on how many handshakes or campaign contributions away you are from persons in power in Washington, D.C.

MAKING AND BREAKING CITIES

Nineteenth-century scientist-inventors such as Faraday and Tesla helped bring the electrical grid into being by zestfully exploring and taming electromagnetic forces. The 20th-century scientist–inventors who harnessed nuclear forces to human goals, people such as Niels Bohr and Robert Oppenheimer, have been more ambivalent about their work, and not only because of the destructive power implicit in the nucleus but also because of the longevity and potency of radioactive waste.

Lewis Mumford, always trying to place our technological aspirations and anxieties in a wider cultural context, compares our search for engineering "fixes" to a recurring theme in folktales, namely the hankering after magic solutions to life's problems:

> When some deep-seated human wish is gratified by magic in these stories, there is usually some fatal catch attached to the gift, which makes it do just the opposite of what is hoped. This catch is already visible in atomic energy. We know how to turn nuclear fission on, but, once we have created a radioactive element, we must wait for nature to turn it off if we cannot use it in a further reaction.[30]

On our Indian Point visit we were as close to the core as possible. Returning to the normal world, we pass back through a gauntlet of detectors. First, a wand is passed over the hands and shoes, the surfaces most likely to pick up stray radioactive residue. Next, we stand in a telephone-booth-sized device that sniffs for radiation over the whole body. After that we surrender the radiation badges, which are scanned for activity. Back out through the security checkpoint, past fences of wire, brick, and concrete. Surrender the identification badges and we're free.

At the end of the day we must drive out of Indian Point beneath the massive transmission lines carrying away the reactor's harvest of electrical power down to New York City. A typical 1,000-megawatt reactor, like Indian Point's, possesses at its heart the fissionable material equivalent of 1,000 nuclear bombs.[31] This reactor, it must be emphasized again, cannot explode because the reactor's fuel is not rich enough in U-235 to make a bomb. Nevertheless, having all that captive energy in one place, quivering at the center of all those nuclei in all those atoms in all those fuel rods, furnishes one last sobering fact. Take the nuclear capability of a 1945-era uranium bomb, multiply it by a thousand, and that's what's being loaded into the Indian Point Unit 2 right now.

And at the same time, 100 meters away, working like a beaver, Unit 3 has been churning out electricity at a mighty clip. Final fission fact: Those transmission lines leading off the property have just carried down to New York City, during the hours we've been at the plant, an electrical equivalent roughly equal to the energy unleashed by the bomb dropped on Hiroshima.[32] Depending on how you configure the forces, you can use uranium to energize or pulverize a city.

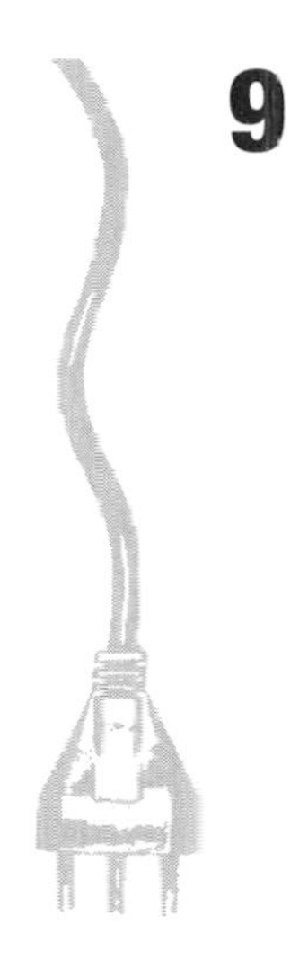

9

TOUCHING THE GRID

Something was wrong. When Mark pressed the button, instead of a reassuring consistent whir there came an ugly grinding gurgle, an uncertain convulsion, an acrid smell, and then silence. After 20 years of service, accelerating fruit to high velocity to make pulpy beverages, the blender seems to be at an end.

Mark performs a tabletop autopsy. He undoes a few screws, revealing the inner workings. The main viscera are two coils of wire, one stationary and one that rotates with the central shaft. A spindly wire forming part of one of the coils probably developed a frayed insulating sheath. All those cycles of heating up and cooling down have led to weakening and corrosion. Some of the electricity tried to take a shortcut, too much current built up, and the internal fuse blew.

Big deal. The world does not stop. An avalanche does not begin. Thirty million people in nine states do not lose power. If you had a sure enough hand, you might try to repair the wiring. In the present network of quick manufacture, chain-store retailing, and cheap electricity, there is little motivation for fixing or even recycling the old appliance, so it gets tossed on the rubbish heap. Mark could actually reach in, grab the wires, and repair the thing—not many are able to do that—but won't since it isn't worth his time. Indeed, he could probably do without the blender altogether, but he'll get another one anyway because it's part of his customary kitchen routine. It represents a bit of marginal convenience. He owns a blender because he can.

I'm at the home of my brother, Mark Schewe. His appliance burned out, but it could just as easily have been your blender. In this chapter we're going to look at how his home, or your home, gets electricity. We'll explore his grid, from top to bottom, sometimes on hands and knees.

THE PARALLEL LIVES OF APPLIANCES

Henry David Thoreau wrote a book about a journey he took with his brother. *A Week on the Concord and Merrimack Rivers* describes plenty of fascinating sights, but the narrative was really devoted to Thoreau worrying about the political, economic, and cultural forces of his day. Like Thoreau's excursion, this chapter will feature a sort of float trip on or near two waterways, the Boise and Snake rivers. Like Thoreau, I shall narrate the sights and examine how technology shapes our lives.

This journey through the heart of our electrified world begins in Mark's kitchen. The blender is blasted, but plenty of other electrical creatures continue to drink at the trough. Powered machines need electricity the way people need food or cars need gasoline. In the case of food, it's easy to see the source. You pick up an apple and eat. You pull your car abreast of a pump and load the volatile liquid into your tank, many gallons at a time. With electricity it's a bit harder to see where the energy comes from. The spigot for electricity—a wall socket—is visible but not the energy itself.

For the grid, paternity is hard to prove. As with New England, Idaho is plugged into a vast internal energy sea through a million outlets. For Mark's bread toaster, did energy come from the giant Brownlee Dam on the Oregon–Idaho border or from the Jim Bridger coal-fired steam-electric plant all the way over in Wyoming or from one of Idaho Power's many dams? You can't tell because the utility makes and mingles energy from all of these sources. It's like drawing water from the Mississippi River at New Orleans; the water might be coming from Cincinnati or Minneapolis or Tulsa or Bismarck. The service area for Idaho Power, some 20,000 square miles, isn't as big as the Mississippi watershed, but it is big enough to render anonymous the electrical power arriving at the socket the instant you turn on that blender.

Some engineers would like to send information over the same lines used for power. So from the same socket, or one like it, you might one day get information services or even computer power. In grid computing you would need only input and output equipment since all heavy duty crunching of data or picture manipulation would be left to some remote processor. One socket would be for generic electricity, another for generic computicity. What next? Thoreau would ask. Perhaps the grid will also think for you. Plug your mind into the wall and out comes thinkicity.

In Idaho a hundred years ago electricity was given out during the dusk-to-midnight hours, with the exception of three hours on Tuesday afternoon so that women could operate electric irons.[1] Those days are long gone, and Mark can count on the fact that electricity is on duty on through the night. He won't have to pump water for his morning scrub or set fire to wood for heat or put match to candle for light. With grid-based amenities there are no fumes, no throttle, no pull cord or primer, no ignition, no general exertion at all except to throw a switch or select a speed setting on the side of the appliance.

In the home all the circuits form a confederation. Each room, or couple of rooms, gets a circuit separate from the others. When you walk down the hall you pass from one electrical jurisdiction into another. All these circuits extend their wires down through the walls. They meet in parliamentary assembly at the circuit (or fuse) box in the garage or basement. There the home's electrical allotment is doled out; the total budget cap in your home is typically 200 amps.

What's an amp? So far in this book I've been talking about electricity in terms of *power* (the amount of energy delivered per second) or *voltage* (essentially the force under which the electricity is supplied). No less useful in describing the river aspect of electricity is the concept of *current*, the number of electrons passing down a wire per second. The current is related to voltage and power in this way—power is just voltage times current. Or to swing things around another way, current is power divided by voltage. All these units are named for various inductees into the electrical hall of fame, a sort of lifetime achievement award. In addition to amps (the unit for current, named for Jean Jacques Ampere), the list includes watts (James Watt), volts (Allessandro Volta), and ohms, the unit of resistance, or the amount of electrical drag imposed by the appliance (and named for Georg Ohm).

Under the constitutional rules of this parliament, individual room circuits are strictly apportioned. Each bedroom, for example, may appropriate no more than 15 amps of current. Let's turn this nebulous amount into something you can *see.* A 100-watt bulb is pretty standard. You might be reading by one right now. To determine the current flowing through the bulb at this moment, just divide the power (100 watts) by the voltage (120 volts); 100 divided by 120 is a little less than 1 amp. In other words, an amp is the current of electricity flowing through a 100-watt bulb. Know your electricity. And it *is* your electricity; these are not electrons coming in from the grid, but electrons there in your wires to begin with. What you buy from the utility is the electric force to drive the electrons already present in the circuits within your appliances. The lamp drinks only a small part of the room's quota of current. You could burn a dozen or more 100-watt lamps and still get away with it.

Some appliances are more equal than others. They are so power hungry that they get a special dedicated circuit all to themselves. These exceptional machines include those that produce lots of torque (turning power) in compressors dedicated to lowering the temperature (air conditioners and refrigerators) and those that raise the temperature (ovens, water heaters, clothes dryers). Upper limits on allowable currents for these monsters can be 20, 40, or even 80 amps. All these appliances, strung out on their separate circuits, with their separate wiring running behind the wall boards, know nothing of each other's existence. Their parallel performance only becomes problematic when the overall maximum current is exceeded.

The rules are enforced with fuses or circuit breakers. They interrupt the flow if it gets too high, leaving you in the dark. The reason for the cutoff isn't stinginess on the part of the utility. The utility loves to sell you power. No, the reason for limits is the continued well-being of your circuits. With too much current, the wires inside your bedroom wall might have melted or started a fire. Much better for an inexpensive, plum-sized fuse to fail or a breaker to flip open and for you to lose power in a room than for you to lose the whole house. Once you've corrected the problem, usually by turning something off or down, you can replace the fuse, or reset the breaker, and you're in business again. The power resumes its regular course.

But should it? Couldn't we turn some of the machines off or down on a permanent basis? If you were to suggest to the homeowner that the melting of the Greenland icesheet could be tied to *his* use of a blender or air conditioner, he might not at first see the connection. Or if he did recognize the link between appliance, burning coal, and climate modification, he would ask, not unreasonably, why someone in Washington wasn't doing anything about it.

Henry Thoreau, who never heard of greenhouse warming, had other reservations. "A man is rich in proportion to the number of things which he can afford to let alone," he said, meaning basically that we should turn off the machine and go read a book. Considering the cost and worry of upkeep, we might discover that we no longer own our machines but rather that they own us.[2]

Mark shrugs. He hasn't time to think about who owns whom. He has to get to work and the clock indicates he's running 10 minutes behind schedule. Thoreau would have smiled at this tiny example of time tyranny.

METAMORPHOSIS

Mark works for Idaho Power, the very utility that energizes his home. He has a specific job—testing and repairing power equipment of all sorts—but he seems cheerfully ready to do all the other jobs too. He knows everyone at work, what they do, what's in their tool drawers, and he has informed opinions on how all the procedures should be carried out.

It's only 6:30 a.m. but Idaho Power's Boise Operations Center is filled with activity. A lot of the employees come in early in order to leave early, or else to work long hours four days a week in order to earn a three-day weekend. The center is a sort of hospital for ailing grid components. The particular patient whose case history we will now examine normally resides about 30 feet off the ground. It's the thing that looks like a garbage can bolted to the side of the power pole. In the Idaho Power system there are about 100,000 of them. This device, a transformer, converts electricity from the industrial scale of 7,200 volts down to the domestic scale of 120 volts. The circuit panel in your basement or the meter on the wall outside your home might be the formal boundary between the utility and the customer, but the changeover really occurs up that pole and inside that can.

These drop-down transformers, as they are called, are the off ramps of the electrical highway system. To know them is to know how your home downloads energy from the grid.

It's only midmorning and already the air conditioners are coming on, and that means a full workout ahead for the utility. Among the sturdiest of strands in the knitted fabric of the grid, transformers are built to brave out the elements and heavy electrical traffic for decades. Some are sexagenarians and still lead productive lives. They are hardy but not eternal, and eventually they succumb to one sickness or another and have to be taken down and brought in for repair.

Let us designate transformers, in their capacity as ambassador between customer and company, as the official representative component of the grid. And to see how they are maintained in good working order, we will follow one particular transformer on its journey to the doctor for a checkup. Just as this chapter traces a day in the life of a single utility, so this section traces a day in the life of a single transformer. Let's call it JK67.

Here at the operations center, JK67, like any patient at a prim medical clinic, will get the full treatment. First comes the preliminary inspection; then a flow of documentation starts up. Date of birth? Time of last checkup? Obvious external damage? Leakage from orifices? Internal bleeding? JK67 is bigger than a blender but surprisingly has fewer parts. What you see when you peer inside are two coils of wire wrapped around a common steel yoke, and that's just about all. Nothing moves. Only energy moves and is transformed, and this is the whole point and destiny of the grid.

To see how the metamorphosis works, please compare the transformer with our planetary system. The moon exerts a gravitational effect on the earth: It induces tides of water to move up and into bays and rivers. In the transformer, currents flowing in one coil exert an electromagnetic force on the other coil. It will, courtesy of the Faraday effect, induce currents to move. The moon and earth never touch; the force between them moves through empty space. Likewise, the two currents in the two coils never mingle; the force between them moves through the steel yoke. The current in the primary coil gives up some energy in order to induce current in the secondary coil, a reciprocal action whose technological development must rival in importance the invention of the wheel and the discovery of penicillin.

The electromagnetic action in the secondary coil, a tiny tide more periodic than any lunar tide of river water, is what proceeds out of the transformer, down the pole, into the service panel, and on into those parallel circuits inside your home assigned to all the blenders, bulbs, and dehumidifiers.

What you buy from the grid is power, energy per time. Technically, you're not buying a current—a flow of electrons—but voltage. The electrons flowing in your wires were there to begin with. Because of the alternating current effect, the electrons are jerked left and then right 60 times a second. They hardly have time to go far, maybe a few inches. They don't so much flow as oscillate. What you're buying is a sort of zest across two terminals, a forcefulness, an agitation that jiggles the electrons.

Why not just take the electricity directly from the grid? Why pass it through a can perched up a pole? Because the high voltage good for *transporting* power is bad for *using* power inside home appliances. The power equals the voltage times the current. You can send the same power by making the voltage high and current low or the voltage low and current high. It's like being given a million dollars. You can get it in the form of ten $100,000 bills—good for transport since it easily fits in your wallet—or you can get it in the form of a hundred thousand $10 bills—easier to spend in a store but more difficult to carry.

"Smell," says Mark as we stand in front of JK67. It looks like olive oil but isn't. I would like to touch it, but I don't. The inside of a transformer is a swimming pool, a miniature spa for electricity. The coils soak peacefully in a bath of about 10 to 15 gallons of oil, a grade of oil not very different from motor oil. The oil keeps the transformer from being French fried. It's there partly to cool the coils, which get warm carrying all that energy, and partly to insulate the coil wires from each other. If two wires touch or come too close to any metallic surface, a short circuit will result. When electrical current takes such a shortcut—when it flows outside its allowed path—bad things happen. Electrical shortcuts can lead to a blender breaking down, a home burning up, or an enormous multistate regional grid turning off.

JK67 shows signs of exactly this sort of short-circuit. Dark smudges on the walls tell us that sparks must have erupted unseen, intermittently lighting up the transformer's interior like an x-ray scan. If electricity is a form of bottled lightning, here is an example of the bottle (one small corner of it) developing a leak. There is in this case also a report from the chemistry lab testing positive for acetylene, a ring-shaped molecule made in the oil when electrical arcing—tiny thunderbolts tracing the path of the shortcut—break out like a disease in the transformer's innards. The chemistry analysis also looks for another troublemaker, PCB, a very complicated molecule that used to be added to the oil to make it a better insulator. But when PCB became known as a cancer-causing agent, it stopped being used. The findings are good: JK67 does not carry the dreaded PCB.

Like the human body, JK67 can get sick in a number of ways. Its brass fittings, the electrodes where electricity enters and departs from the can, get tarnished and then don't conduct as well. The gaskets, the rubbery layer that forms the snug fit between the can and its lid, sometimes crack, giving oil a chance to escape. Less oil means less insulation and less cooling. That can lead to arcing.

Bushings can be faulty. A bushing is the bell-shaped ceramic cowling that girdles the power line where it enters the transformer. Stop now to consider the bushing's role in maintaining the around-the-clock flow of electricity. Half of the power industry is devoted to making and sending electricity, the other half to restricting that electricity to its proper path. The bushing's job is to keep the lightning bottled. It must at all costs prevent the incoming power from taking a shortcut by leaping from its own wire at 7,200 volts over to the shell of the transformer itself. Electricity, as you have already seen in a number of ways, is an exuberant conveyance of energy and like a frolicsome child playing a game will often make up its own rules. It will gleefully seek out the shortest, easiest path for itself. Wherever this arcing temptation is greatest, there the engineer must insert insulation. Insulation, the opposition of conduction, is designed to frustrate electricity into behaving itself.

In this respect the bushings stacked up on top of transformers, or power lines, or switches (all places where surfaces at very different voltage are in dangerous proximity) are a vital means of keeping electricity in its groove. If the electrical grid were a gothic cathedral, bushings would be its flying buttresses. They have a flaring geometry, are positioned on top of or next to each other in functional orderliness, are stationed on the flank of the main architectural element (in this case the power-carrying cable), and in general support the work of the grid with sturdy yeoman service.

Bushings are built to last decades, but exposure to sunlight can cause them to wither and crack. Indeed solar ultraviolet is harmful to human flesh, ceramic bushings, and many other organic compounded materials. A glaring day like today, a delight to recreationalists, is a detriment to the grid. If you want to make your plastic lawn furniture last longer, Mark advises, rub it with sunblock lotion.

Like a surgeon, the technician plunges in. JK67's short-circuit problem is soon mended. The oil is replenished. The old oil becomes an additive in explosives or, if it is too far degraded, sent to a place in Utah to be burned. Gaskets, bushings, and fittings are replaced as needed. Then the can is closed up and tightened down. Rust spots are sanded off and new paint is applied. The unit is given a road test, in which it is forced to carry twice the maximum current handled in ordinary service. JK67 is labeled, given one last inspection, and then staged on the loading dock where it waits to be taken out to active duty as if it were a pet at the pound waiting for loving owners to show up and give it a home.

In its previous tour of duty JK67 might have delivered electricity to a family on through its childrearing years, a number of graduations, a wedding, a funeral, and war in Vietnam. Very soon a line crew will take it out and hoist its 500 pounds up to the top of a pole. Returned to active work, it will spy on grazing cows and pickup trucks and will deliver enough energy to power many holiday dinners yet to come. In all likelihood no human eyes will peer into its oily depths again for a half-century or more.

THEY ALL HUNT ELK

Jim Terrell is one of the people closest to the grid. It is fair to say closest since he and his fellow linemen literally *touch* electricity. The linemen go wherever the line goes over the geographic entirety of the system. This can mean a conduit beneath a suburban shopping mall or, since there is plenty of desert and forest in the state, it can mean a wolf-inhabited ridge or forlorn alkaline gulch. Someday, if electric power is beamed through the air as waves, the companies will use wave men. But as long as power is sent across bulk metallic lines, they'll need linemen.

It is to a dry terrain of sandy soil we go today because of a report of a warped cross arm on a power pole. This particular imperfection was spotted by a circuit-riding engineer referred to as the Troubleman, an inspector who does nothing but find fault. He looks at lines, poles, breakers, and all the other physical sinew and organs of the grid. Like the tax collector touring ancient empires, the Troubleman comes to each province of the grid only about once in 10 years. He makes spot judgments. Will this equipment perform well for the next decade? If not, then it needs tending.

Jim Terrell is the foreman and leads our three-vehicle convoy down Interstate 84, along a narrow blacktop secondary road, and finally out onto a gravel route winding through quiet rangeland. He is swarthy from long afternoons like this one spent under the solar ultraviolet. Today's job is routine, he says, nothing compared to some of the emergency tasks his crew is asked to perform. Generally held in reserve for first-responder duty in the toughest cases (like the Marines: first in and last out), Terrell's team does nonemergency work as the schedule permits.

The trucks have reconnoitered at the appointed spot and pulled off onto the shoulder. Blueprints are spread out on the dashboard. Even from the road it's easy to make out the specific crossarm in need of help. You can see the warping—it looks like the curved yoke used for driving oxen, not the straight-across wooden support it's supposed to be. Wouldn't last 2 years, much less 10.

Mr. Terrell, with 34 years on the job, is a jack of many trades, including the maintenance of good public relations. He could claim the company's right-of-way ability to cut the fence that protects the property and then just charge up the hill like Teddy Roosevelt and manhandle the pole. But diplomacy is always good policy and, walking up past a battery of barking dogs and suspicious looks from inside the house, Terrell politely asks the homeowner if she would mind, please, if the power company proceeded with its necessary repairs. Permission is granted.

Actually, the crew has been out this way before along the same road. While doing their work on that occasion some other men had driven up, stopped not more than 100 yards away, and set up for a bit of target practice. So here they were, the linemen doing their job handling high-voltage equipment, and gunfire was going off nearby. This led to an exchange of words. The men of the work crew weren't against shooting. In fact, they all hunt elk. This is the West. On weekends, they go off the grid and up into the mountains. Like Mark, three of the four linemen even use bow and arrow to make the pursuit of game more sporting. But on that day the sound of bullets flying was disconcerting to the fingertip-control needed for attending to 7,200 volts, and so the target shooters were politely asked if they wouldn't mind moving farther off, which they did.

No gunfire today, and things proceed speedily. The men carefully pull down a few fence posts, snip some barbed wire, and then drive off the road and move jarringly up an uncomfortable incline. These aren't compact off-road vehicles. The flagship, a rig about the size of one of those towtrucks used to rescue buses in distress, is called a bucket truck since its main accouterment is a crane that lifts two men, each in his own steerable pod. For the next three hours it will serve as their elevated workshop.

The bucket truck is parked at an odd angle with respect to the hill but then deploys retractable struts that brace the vehicle billy-goat style, hoisting it off its own back tires. Once the vehicles are settled, the men deploy. The designated twosome climb into the movable gantry with their specialized gear and start moving upward, maneuvering the buckets in three independent directions upward to within inches of their quarry. Now truly they are at eye level with the grid, whose lines stretch out laterally for miles behind and ahead. Big cities might bury many of their transmission lines, but most of the world grid is up *here*, 30 feet in the air—a couple of tons of metal wire per mile—-held aloft by the weight-lifting exertions of utility poles and their affiliated limbs.

The job is straightforward: Affix a new arm to the pole, transfer the power lines over, and then remove the defective arm. The men riding the buckets, in their protective gear look like astronauts, but this is not a space mission. They're at an altitude of only 30 feet, not 100 miles. Their forward velocity is zero, not 17,000 mph, and they do not confront the hazard of interplanetary vacuum. On the other hand, the linemen do have to worry about something the astronauts do not: the hair-raising presence of 7,200 volts.

It's high noon, and today's work is being done hot. That is, the voltage has not been turned off. There will be no inconvenient interruption of service, and current will continue to flow along the uninsulated cables during the procedure. Sometimes the linemen use a "hot stick," a hockey-stick-sized insulated pole with a controllable hook at one end, for close-order gaffing and manipulation. You'd think this would be like doing precision jewelry settings with a pair of chopsticks, but actually a practitioner can get very good with it and accomplish many intricate repairs.

This time they're not using sticks, but gloves. With rubber gloves, rubber arm protectors up to the shoulder, and rubber blankets draped over the surrounding surfaces, the line can be touched. Now, this touching of a fully energized high-voltage line does not come naturally. A cavalry steed will not gladly jump a hedge or charge an opposing foot soldier. It goes against the horse's instincts to run into things. Going against instincts can, however, be attained with sufficient training. The horse can do it and so can the man. The man does not gladly grab a hot line, but he can be trained to overcome this particular dread.

The company has a special school for linemen. At this electrical West Point the rookies shimmy up tall poles and are confronted with some of the grid's sternest lessons. Training, the passing on of technical knowledge with life-or-death implications, is a crucial part of the present activity. Therefore, to further the educational process on today's outing, seniority will be turned on its head. Thus, the man with the least experience, an apprentice lineman with only six years on the job, will take the bigger share of the work. In the other bucket only 2 feet away, and performing the second most important tasks, is the man with the second least amount of experience, a mere dozen years. Next up in seniority and next down in order of importance to the immediate proceedings—indeed he does the ground-level grunt work supporting the men above—is a lineman with a quarter-century of work behind him. He is always looking up, checking, fastidiously assuring himself that the safety procedures are being honored.

Finally, there is Jim Terrell, who orbits the scene at an even greater radius and perspective. He broods over the whole production, circling around the pole and truck like a border collie, saying little but recording everything. He insists that the apprentice be foremost in the work. He wants the youngster above to think through the issues, especially the 7,200 volts, and to do the whole job. How else do you learn?

Asked about the frequent danger, Jim shrugs. You can't afford to let your concentration wander. The 7,200-volt current is always there waiting for its chance to jump to a new surface. Hundreds and thousands of repairs at 7,200 volts have been carried out with no ill effect. But then there is the case of their unfortunate fellow lineman, and perhaps the men are thinking of this painful event as they try to finish their chores.

The recent mishap would remind them, if anyone needed reminding, of the power in their hands. In their elevated position high by the crossbar beneath the noonday sun, the linemen are the very priests of the gods of electricity. Wearing the appointed vestments and gingerly arranging the special implements around their electrical altar, they are the individuals most intimately close to the thunderbolt. This electromagnetic bottled lightning streaking about in wires near light speed is a force of nature. Yet it is also obviously unnatural since it is patently manufactured by humans for their convenience and pleasure. The advantage of domesticating lightning is so great that society pays these four crewmen extra hazard compensation for voluntarily grappling with the ragged, lethal edges of the grid's intrinsic ferociousness.

The men still up in the buckets at the 7,200-volt level know that the guy was released just a few days ago from the hospital. What happened to their colleague is by now well known, at least within the company. The incident took place not up a pole but down in the dirt. He reached out for a line at 120 volts with gloves rated up to 600 volts. He should have first tested the wire with his meter but didn't. Always measure the status of an unknown line is the policy. Kneeling down close to his work, what he found in his grasp was not a wire at 120 volts—the wire on the home side of a transformer—but one on the industrial-strength side. His wire was at 7,200 volts.

This was full, undiluted bottled lightning, more than enough energy to send you from here to eternity. The bolt, ever eager to find a shortcut, instantly went up the man's arm, through his body, and out the two points closest to earth, the knee of one leg and the foot of the other. He was flaming so much that his nearby work mate emptied a whole fire extinguisher putting him out. A burn like this is beyond burn. It is a roasting from within. It's an intensive laying on of external energy to an internal place it should not be. The ordeal was horrific. He was in a coma for three weeks. He should have died, but he came through. The treatment was not pretty, his heart is forever weakened, but he's alive.

Nothing remotely resembling this happens on today's outing. Like surgeons sewing up a patient after a successful transplant, the grid, at least this small part of it, has been restored to health. The men retract the tools and shrouds and make their final check of wires, bushings, and bolts. Gantry down, paraphernalia stowed, refuse picked up out of the dust, and the job is done. As if seeing a movie of their work shown in reverse, we see the trucks back down off the hill, fence posts go back up, and barbed wire is reattached. Everything is neat and professional. No loose ends are left behind. You'd hardly know that 73 years' worth of know-how had been allocated here for several hours sprucing up the grid.

The greatest investment by the company is not in buildings or equipment but people. Those 73 years of collective experience clearly outrank all other grid figures of merit. The wires carrying the electricity will last 15 to 20 years, which is longer than the expected lifetime of an elk in the state of Idaho. The crossarm just installed with the new bracing system and more long-lasting resin bushings should go 30 to 40 years. The pole, even with its tarred base sunk in the dirt where moisture can get at it, will last you some 50 to 60 years. This is impressive, but it can't match the longevity of a lineman—80 years or more—who's been careful at his work.

A TIGHT FIT

The grid in Idaho lies mainly in the plain. That's where the people are and where the water is. The Snake River, starting east in the Teton Mountains of Wyoming, flows in a graceful arc across the dry, flat southern portion of the state before turning toward the mountains in the north and on into Oregon to the west. A large river provides lots of things: water for irrigation, a picnic destination, inspiration for paintings and photographs, transportation, fish, and, since the coming of turbines, a source of power. Most electrical energy comes ultimately from the sun. Sunlight from hundreds of millions of years ago, invested in coal, reemerges in a steam engine when coal is burned. Sunlight is invested in evaporating water from off the vast plain of the Pacific Ocean, water that wafts inland, collides with mountains, and tumbles down the gravity gradient back toward the sea. Some of this energy reappears as electricity when the water passes through hydroelectric barriers thrown across rivers.

This is where the poetically minded historian could slot the story of the grid into the much larger pageant of cycles filling out our natural environment. Again I take inspiration from Henry David Thoreau, who could see in ordinary activities, such as tending his garden at Walden Pond, an action just as poetic and profound as the greatest deeds of the Hindus, Persians, Babylonians, and Egyptians.[3] Is the creation and maintenance of an electrical network likewise a poetic deed? I guess if

I didn't believe this to be at least partly true, I wouldn't be writing this history of the grid.

The utility in Idaho came into existence to facilitate the last stage of the conversion of solar radiance into a form that abets earnest human activity. In the beginning there was Swan Falls, the place where the company that became Idaho Power built its first serious dam more than a century ago. The structure rests within a steep canyon in a beautifully forlorn stretch of the Snake River south of Boise. The only road leading in passes through the heart of the Snake River Birds of Prey Natural Area. Mark has in the past kept a predatory bird or two himself as a pet and can look at a spot on the sky from a distance of half a mile and tell one species of hawk from the next.

When we arrive in the warming part of the afternoon, the men running the installation are outside on the dam itself doing the daily dredge. Before Swan Falls can harvest electricity from the passing water it must first harvest grass and other debris snagged on the filter guarding the intake ports. The stuff comes out in dripping bales and is taken away by truck. This must be one of the more homely tasks of maintaining the grid, and you wouldn't say that the operation was grand on the level of the Egyptian pyramids. In counterargument you could fairly say that the pyramids, spectacular as they are, never generated any electricity.

The original turbines at Swan Falls, constructed not long after those at Niagara Falls, have been retired, and the old powerhouse, still sitting on top of the dam wall the way Tudor homes used to perch on Tower Bridge in London, is now a relic of the days when gold prospecting was still a big reason for coming to Idaho. In fact the powerhouse has become a museum where schoolkids can see the big machines and photos of linemen now long dead. Actually, the museum itself has now become a relic since security concerns have recently caused the place to be closed to all but the occasional special visit. Unpadlocked for an hour, the main hall seems, like many other exhibits of obsolete heavy machinery, to be a shrine to past power. The old dynamo is still bright green and looks as if it could work again if asked. It gives out a faint smell of lubrication, a reminder that much of the mechanical world skates on a thin layer of grease.

This being the end of summer, the annual reserve of Teton snow is now much depleted, and so the melt-fed river flux is impoverished. Without enough gainful employment, one turbine takes a rest and its mate shoulders all the work. When snow is down, the river is down, and consequently power is down. The plant pretty much runs itself. The few personnel on duty get regular computer diagnostics on what is happening. The dam speaks to them. Dozens of indicators provide water facts, such as river level and wind speed (wind can push water up against the dam), and electrical facts, such as the amount of power being produced at Swan right now. Besides the trim banks of dials, switches, and lights, what strikes a visitor in the byways of the building is the color-coded piping. Pipes run everywhere, hinting at the kind of movements quickening all around: orange pipes for oil (at a dam you need hydraulic pumps for opening and closing things), green for compressed air (for actuating more pumps and switches), blue for cooling water (anything carrying power gets warm), and red for fire water (that is, water for extinguishing fire).

The business end of this energy factory is the turbine, the contraption through which the entire Snake River squeezes onward by passing down a tunnel and being forced to dash itself into a huge pinwheel, which in turn bequeaths its rotation to the generator apparatus for cranking out electricity. The process is pure aerodynamics—all fins and flaps and streamlines. The angle between the turbine blades and the oncoming aqueous whoosh can be pitched to vary the torque, and in this way the desired power can be obtained. In effect the turbine is forever swimming against the whole of the river without ever getting anywhere.

It's all in a good cause, this swimathon. With the watery husk of the river spiraling out downstream of the dam, the sweetmeat, the protein part of the Snake—water flow turned into shaft rotation turned into electrical current—is sent up to the powerhouse roof where it energizes three thick cables in the form of 7,000-volt electricity, quickly jacked up to 138,000 volts in a grander version of the transformer on the pole next to your home. Given its new suit of clothes for long-distance traveling, the Swan Falls electricity runs through its wires, up the steep canyon wall, makes a straight-arrow sprint to join the general grid, and might, after nary a second's delay, make an appearance in Mark's blender.

Would the customer know that the energy for spinning his blender had been, only a millisecond before, an organic part of the Snake River, 40 miles away? Not likely. All over the map tiny pieces of the river were being peeled off and teleported to distant appliances without any fuss being made at the audacity of it all. Our awareness of the process, if there had been any, would be dulled into oblivion by familiarity. Thoreau was ever the keen observer of the background processes that fill our lives: "The longer the lever, the less perceptible the motion. It is the slowest pulsation which is the most vital."[4]

Hyroplants like Swan Falls consume no costly fuel, spew no carbon compounds that defrost the Arctic ice pack, and form no acidic sulfur compounds that slay trees 500 miles downwind. To meet new energy needs and keep all the blenders whirring, why not just build more Swans? The short answer: damnation. Or not enough of it. Or maybe too much of it already. In North America and Europe and other built-up places, most of the suitable hydro locations are already taken. Furthermore, some of the existing dams have proven to be mistakes. They collapsed or silted up too quickly or sacrificed too much topsoil. Some (for example, in the Florida Everglades) will have to be dismantled or altered since they have messed disastrously with the food chain, that universal and complicated woven linkage among living organisms in which Homo sapiens are but one part. A dam is not taken down for sentimental reasons. Even when authorities aren't particularly concerned with the extirpation of some exotic species of minnow, they do have to pay attention when the sport fish that fed on a species that fed on the minnow becomes endangered.

Like the inhabitants of the Tennessee River Valley, the citizens of the Snake River Valley also have become greater in number, more avid in their enthusiasm for powered machinery, and more assiduous in their pursuit of hearty agribusiness, and so new forms of power generation were needed when the electrical potential of the river became fully booked. With more mouths to feed and more blenders to spin both of these river region grids founded on hydropower, the TVA and Idaho Power, were forced to turn to fossil fuel.

SQUEEZE AND BANG

Gary Felton, a friend and hunting partner of Mark's, is explaining the game of American baseball to some technicians from the staff of Siemens–Westinghouse, the giant manufacturer formed by the amalgamation of two historicallly significant electrical companies (later bought out by a British company and, last year, by Toshiba). What are these Germans, forced to wear sunglasses to keep out the awful glare reflecting from this parched cowboy landscape, doing so far from Europe? They're here because they make just the perfect machine for the job at hand. They're 7,000 miles from home in order to make sure everything continues to go well.

To see how the combustion side of the company functions, we've now gone from Idaho Power's oldest installation, the hydro plant at Swan Falls, to their newest installation, the Danskin gas turbine unit sitting near the town of Mountain Home. Here the fuel is natural gas, delivered to within a few hundred meters of the fence by a pipe network that stretches all the way to Texas. Like the gas itself, the gas grid is invisible. Unlike the electrical grid that lopes along for miles out in plain view, gas pipes often lie beneath the landscape. For security reasons, gas merchants prefer anonymity.

What you do with the gas is shoot it into a chamber where it meets with a blast of air compressed in a two-story structure that, although it lacks a Mercedes hood ornament, looks like the huge front grille of an automobile. This is appropriate since the whole operation gives you a sense of automotives or jet aviation. The fuel and air meet and produce a potent combustive blast that spins a turbine connected to a generator. They don't bother making steam; the exploding gas itself makes the turbine go. Once the gas has done its job, it zooms out a chute and up a stack.

Gary, the operator on duty, reports that the familiar names for these crucial sequential steps are *suck, squeeze, bang,* and *blow.* What you've got, Gary says, is a jet engine that goes with an oomph of 60,000 horsepower. Instead of applying this turbo energy to sustaining transatlantic flight, a pouncing load of high-voltage electricity is produced. The Danskin facility will never rival the historical importance of Jumbo Number 9, or Red October, or Faraday's coils. Instead, its great virtue is its very anonymity. The turbines are anything but custom built; they were made and assembled quickly.

I ask Gary why they aren't using the more efficient type of dual-cycle gas turbine that uses the waste heat to make extra electricity. He shrugs and reminds me that this turbine's mission is to help the grid get through late-day surges in electrical demand. Efficiency is secondary to that goal.

Here at 3 p.m. only one of Danskin's two units is busy, so we can peek inside. Normally it's too hot and noisy to be in the generator room. Gary encourages us to reach under the covering and touch the drive shaft, the rotor usually spinning at 3,600 revolutions per minute by the force of the gas combustion hitting the turbine only a few feet away. Although it's not producing power today, the shaft is kept slowly rotating anyway so as not to deform as it cools down. So, here I am, not exactly holding a high-voltage line in my hand, but I am holding, ever so gently, the revolving central part (mechanized but fortunately not electrified) of the grid. The machine offers its underbelly to be petted. At one end of the shaft is the turbine, which feels the heat of the gas. At the other end is the rotor-mounted coil, which feels the load of all those blenders. Thus a ribbon of force connects puffs of gas coming out of the ground in Texas with ice cubes being crushed in Idaho.

Do we make too many ice cubes? Do our social engines run at full throttle too often, thus compromising some of the simple pleasures in life? Thoreau thought so. He believed we live too quickly.

> Men think that it is essential that the nation have commerce, and export ice, and talk through a telegraph, and ride thirty miles an hour. . . . We are in a great haste to construct a magnetic telegraph from Maine to Texas; but Maine and Texas, it may be, have nothing important to communicate.[5]

Indeed, in the century and a half since Thoreau's time, we possess even more labor-saving devices and conveniences far beyond the scope of telegraphy. Most of these machines tend to make us live life even faster than before. And having said that, I must hasten off to the next appointment.

VOLTS AMPS WATTS

Once the electricity rises up out of the generators, how does it come the final mile to where you live? First, remember that electrical current isn't sent to your home. Instead what's happening is that a certain standard voltage is maintained where power is used. You have your own electrons; the utility provides volts. To keep voltage up, to replenish the energy siphoned off by you and your appliances, someone somewhere has to keep refilling the bowl so that it's always brimming.

Marsha Leese is responsible for this. She is chief power dispatcher for Idaho Power. When asked if it's unusual for a woman to hold such a job, Marsha shrugs. (Everyone in Idaho seems to shrug.)No, she says, it's nothing special. Marsha's boss is a woman. Marsha is at the fulcrum. On the one side are generators subsisting on fuel in gaseous, liquid, and solid forms—methane, hydro, coal—and on the other side of the supply/demand equation are the teeming mass of customers. She tells the generators how much to make and when. She is the managing editor of the grid, with the solemn duty of perpetually keeping the lights on in Idaho.

As such, Marsha helps to landscape the energy environment of your life without your knowing it. Thoreau has this to say about such silent officiators:

> It is something to be able to paint a particular picture or to carve a statue, and to make a few objects beautiful; but it is far more glorious to carve and paint the very atmosphere and medium through which we look, which morally we can do. To affect the quality of the day, that is the highest of arts.[6]

Marsha administers all the various voltage realms. At the top end is the level of hundreds of thousands of volts. In this major league the players are the generators themselves and the largest of transmission lines, the ones stretching hundreds of miles and leading out of state or leading to Idaho Power's main switchyard. The next level down is 69,000 volts, used in taking power from the main switchyard to the substations in the towns and large suburban jurisdictions making up the company's customer base. Each substation has a minor switchyard of its own, where the voltage is lowered and the power divided further. At this point we go from the town level to the neighborhood level. Coming out of the substation is power at 7,200 volts running along lines that typically travel a few miles or less. The final editing consists of stepping down the voltage to a domesticated 120 volts at individual homes. All the power divisions and redivisions and voltage conversions described here are played out pretty much the same way in your neighborhood.

For a moment I stand at, and resist the temptation to touch, the console where a lot of the dispatch is handled. The operators who perform here might not have the fingertip finesse of Arthur Rubenstein at his concert grand piano, but their keyboard technique is significant. They play for a subscription audience of consumers who hardly give a thought to the concerto of volts coming out of this room except on those rare occasions when it stops.

So it was a few months before when, on a 97-degree day with air conditioners everywhere all set on "Hi Kool," the gods of electricity decided to make things difficult by disrupting the flow along two separate lines transporting power from Idaho Power's dams in Hells Canyon. As the name of the place suggests, the terrain is rugged and the power lines are more likely to be visited by mountain goats than humans in any given week.

There began now a race against the clock. Out-of-state power was being rerouted in, but would it be enough to make up for the loss? After the Hells Canyon line went down, the company issued a request to customers to reduce their demand. But with the missing transmission lines, the climbing mercury, and the oncoming rush hour, things were moving to a head. The Idaho grid, like much of modern technology, is positioned on the steep slope of a mountain of complexity.

As the minutes went by the options narrowed, and it grew apparent that the utility was coming up short in its effort to fill all the circuits with energy. Load shedding, the process of deliberately turning off part of the system in order to save the rest, now presented itself as a distressful possibility. The power dispatcher's instincts tell her to keep the power flowing and not to snatch it away. Her sacred duty always is to keep the lights on. Marsha has never been in the control room at Consolidated Edison, but she knows what it's like to respond to a massive electrical failure as it happens. As part of her own training, she heard tape recordings of the grid dispatchers in New York City during a titanic blackout. She knows how *not* to behave in a crisis.

But on sober reflection the dispatcher remembers an even higher duty, which is to keep the *system* on. Consequently, at around 5:30 p.m. that day off went the lights for about 35,000 customers in several counties. About 170 megawatts of load was disconnected.[7] For three hours no blenders spun in numerous homes, and irrigation in many fields ran dry. On the positive side, the rest of the grid went on performing.

As if to underscore the difficulty—technological, economical, and moral—of overcoming the dispatcher's creed of keeping the lights on, Marsha points to a picture hanging on the wall near the control console. This framed declaration specifies that no dispatcher will be reprimanded for shedding load, providing that all other avenues of matching available power with the instantaneous demand had been tried.

The company does not make public the procedure it uses, when push comes to shove, in sequentially shedding load of varying sizes. But rest assured, there are reasons for everything. These bigger grid decisions are made near the top of the company's headquarters building in downtown Boise.

UPPER MANAGEMENT

James Miller exudes confidence. He has the calm authority and articulation of a candidate running for reelection to public office. Actually, he's not running for political office, but he is perpetually running a campaign, and he helps formulate the party's platform. As vice president of Idaho Power, he has to satisfy several constituencies. For instance, he must make customers happy by providing reliable power at low rates. He must make regulators happy by working within a labyrinth of ever-shifting federal and state rules. He and his fellow officers at the firm must adroitly provide both short-term profits, much beloved by the investor owners of all corporations, and stable long-term growth of the company. Immediate gratification and long-haul planning are like two oxen yoked together, the result being that the cart sometimes lurches to the left, sometimes to the right, but seldom straight. Mr. Miller is one of those who has to drive the team as best he can. He doesn't expect a smooth ride.

Overall the company is in good shape. The rates Idaho Power charges its customers are among the lowest in the nation, and the company was recently awarded a high place on a list of the best managed utilities. The company's annual report is filled with pertinent information, but a more interesting document for grasping the dynamics of the Idaho grid is the Integrated Resource Plan, the utility's grocery list, or wish list, for the next 10 years. Projects to be undertaken, facilities to be built, contracts and licenses to be renewed, risks to be run, expected regulations to be imposed, unavoidable hazards to be faced—they're all here.

And right off underlying assumptions have to be made. Like generals overseeing the national defense, the Idaho grid specialists must prudently allow for the most inconvenient combination of conditions: low flow in the Snake River (less water means less hydro-generated power); higher than usual loads (temperature extremes require more cooling and heating); continuing volatility in the fuel markets (assume security instability abroad and political wrangling at home); and a steady influx of new customers into the service region at a rate of 10,000 per annum.

The plan develops a business strategy for dealing with these future needs. It formulates a dozen rival packages, or portfolios, of recommendations for action. At the end of the report a winner among the portfolios is declared, and this then becomes the 10-year blueprint.[8] What are the big issues?

Congestion. One of the most urgent problems is crowdedness in long-distance interstate power transmission. Everyone recognizes that there isn't enough superhighway to send all the power that needs sending, at least not enough if you want to have some standby emergency-carrying ability—the equivalent of having a wide shoulder and extra lanes for rush hour traffic. And here the company is in a bind.

Restructuring. Idaho Power is a sovereign business, and yet it is subject to the rulings or guidelines of numerous governing bodies. On one side, the Federal Energy Regulatory Commission, or FERC, would like the company to divest itself of its transmission lines—sell them off—in order to increase competition among rival energy companies. On the other side, the Idaho Public Utility Commission, which regulates Idaho Power's activity, views things differently. It's not against encouraging competition. But the commission does want to ensure that state residents get their electricity without fuss. Therefore, it wouldn't mind if the company kept its transmission lines. And that is what the company intends to do.

Not in my backyard. Everyone wants power, but no one wants a transmission line built in sight of her kitchen window. Most would prefer the power lines to be buried in the earth. But that costs more money, and no one wants rates raised to pay for it. For example, one Idaho town was happy that the utility had extended a large transmission line to its vicinity. Having enough power helps the city market itself to potential businesses and homebuyers. The line was built. The utility then wanted to extend the line to the next town west, but the first town said no. We won't allow you to build an ugly line through the center of our town. Okay, the company said, we can go around your city center or we can go underground, but that will cost extra. The town's answer: No, you can't go through our town, and no, we won't pay extra for underground installation.

Demand-side management. Customers can take some matters into their own hands. They can, for example, choose to accept *poorer* service. Reliability in the Idaho Power system is pretty good, maybe too good. Everything has a price, even reliability. For a reduction in their monthly bills, customers could allow their power, or the power flowing to selected machines such as water heaters or irrigation pumps, to be switched off by the utility during an emergency or during times of peak demand.

Real-time metering. Instead of getting a single lump charge each month (the number of kilowatt-hours multiplied by three fixed factors for generation, transmission, and distribution) you would billed at a variable rate—more during times of peak demand and less in off-peak hours. This real-time price signaling would encourage the consumer to moderate her energy uses. You'd run the dishwasher when you go to bed rather than right after dinner knowing that the price of power drops to one-third.

Environment. The combustion products associated with burning fossil fuel produce things that are bad for our lungs (pollution), bad for trees (acid rain), and bad for climate in general (greenhouse warming). The idea that these emissions are a burden on society and should be taxed is now being assimilated into energy planning not just by environmentalists and public health officials but also by legislators and by the energy companies themselves. Idaho Power will take into account the social costs of gaseous emissions. It expects to be taxed. Just as you pay to put your garbage into a truck that comes by once a week, so utilities would have to pay to have their "garbage," their generator emissions, hauled away into the sky. Accordingly, the company might soon be charged as much as $20 or more for every ton of carbon dioxide it sends up the flue.[9]

Cap and trade. Another way of dealing with the carbon-dioxide issue, making it seem more like an ordinary billable expense of doing business, is for the government to declare a cap on emissions. A power plant might emit more than the limit allows but would have to purchase a waiver for that amount from an under-emitting company. The waivers themselves would become valuable commodities to be bought and sold in a market of their own. This carrot-and-stick approach has been successful in reducing sulfur emissions in a big way.

What do the contestant portfolios at Idaho Power say about these issues? Let the competition begin.

PORTFOLIO 11

Do we live in an age of unprecedented social unease or technological change? Almost surely not. All ages are beset with their own problems. In Thoreau's day railroads multiplied 10-fold the distance a person could travel in a day. The telegraph multiplied by 10,000 the distance words could carry. Here is Thoreau again, as if he were contributing to Idaho Power's Integrated Resource Plan, except that he's speaking from an 1840s perspective:

> Facts are being so rapidly added to the sum of human experience that it appears as if the theorizer would always be in arrears, and were doomed to forever arrive at imperfect conclusions.[10]

The Idaho Power planners must be theorizers. They are required to file a 10-year plan. True, their conclusions will be imperfect because knowledge of future events is imperfect. The Heisenberg Uncertainty Principle has not yet been extended to business administration but should be. But there must be a plan.

They must contend with the uncertainty about resource timing. What if a new plant is finished too soon? Then it underperforms and has, in the meantime, tied up capital that might have done more good elsewhere. Or, worse, what if it arrives a year too late? Then the company scrambles to make up the shortfall by purchasing expensive power from out of state. Cost overruns can throw everything off. What about borrowing interest rates? Will they be 7 percent, or 5 percent, or more like 9 percent?

One candiate portfolio suggests that all new electricity would come from the old-fashioned but reliable way, by burning fossil fuels. Coal prices are pretty stable, but, oh, all those emissions. Another portfolio calls for 1,000 megawatts in wind-generated electricity. This proposal is quite green—no emissions, so no extra carbon taxes—but how reliable is the wind? Is it blowing when you want it? The company's working assumption is that for 100 megawatts of wind-harvesting machinery, what you actually get on average would be about one-third of that.

Each portfolio has its own way of grappling with contingency. Will a carbon tax be imposed? Will the state regulators grant a nice rate hike? Will the federal government renew the licenses for operating all those hydroelectric plants? Will some exciting scientific discovery lead to a new energy source we can't dream of now? There are no definite answers.

Repeatedly, the word *risk* appears in the plan. By the end, at the point where the winning portfolio of options is declared, no reader would be under the impression that the 10-year prognostication is a sure thing. Each of the rival plans has certain points that make it robust or vulnerable relative to each of many engineering, environmental, business, and regulatory risks. The only certainty, in addition to death and taxes, is that the future will come. A plan has to be chosen. Electricity has to be delivered around the clock, now and forever.

And the winner is Portfolio No. 11, a plan for all seasons. It embraces a bit of the old—a 500-megawatt coal-fired plant, alas, with no provisions for gasification or carbon capture—and the new—300 megawatts of wind power. The forecast anticipates that a carbon tax of $12 per ton *will* be imposed. In 10 years the company expects to be getting 38 percent of its power from hydro, 48 percent from thermal sources (coal and gas), 9 percent from nonhydro renewable sources, and 4 percent from purchases outside the system. Perhaps the most poignant number is the percentage of power coming from the hydro source. That 38 percent seems modest for a company that was built around energy taken out of the river. This percentage is bound to decline further because no new dams are being built, nature isn't making any more water, and the number of thirsty customers is constantly rising.

There are also ominous signs that the river flow itself is declining, declining not just in the way the river level rises and falls with the season and swings back and forth in multiyear cycles but also in an absolute sense. The aquifer that hides beneath the northwestern states is losing water to irrigation, and the natural forms of groundwater replenishment are not keeping up. In Idaho many years of well withdrawals have driven the water table lower, and this in turn subtracts water that would have found its way into the Snake and then into the turbines of Idaho Power. What state, farming, and business leaders are going to do about this looming problem is unclear.[11]

Jim Miller must prepare not only a 10-year plan but, believe it or not, a 75-year plan. Visualizing the grid, or society itself, that far into the future makes even the most pragmatic engineer into a Thoreauvian philosopher. He is proud of his company's traditional connection to the river running through his state. On the wall he has two photographs. One shows Shoshone Falls on the Snake River. Besides being a wondrous spectacle of nature, Shoshone is the site of an Idaho Power hydroelectric plant. Mr. Miller enjoys the outdoors and loves to take his children hunting. "It's practically a law in Idaho that you have to hunt," he jokes.

The other picture on the wall depicts a ceremony in which company representatives, leaders of the Nez Perce Indians, and government officials signed an agreement securing hunting and fishing rights for the Nez Perce who, after all, were on the scene long before 138-kilovolt electricity ever flowed. And before even the Nez Perce arrived in Idaho, fish had been there in the river. The stressing of fish and the lowering of the water table might be ecological and civic disasters in the making, but in terms of geological time these matters will be of small account. Nature and the river will always trump human technology.

Not surprisingly, Henry David Thoreau was keen on this subject. He always took the long view. In our day the issue is hydroelectric power on the Snake River. In his day it was the textile mills on the Merrimack River in Lowell, Massachusetts, powered by dammed water. These engineering marvels (no longer in existence) made Lowell an economic powerhouse and a huge employer of factory workers. But this human intervention in the landscape ended the free run of fish up the river. Nevertheless, what man had put asunder nature might one day restore. There is a season for everything, Thoreau felt, and the time of the fish might come again:

> Perchance after a few thousands of years, if the fishes will be patient, and pass their summers elsewhere meanwhile, nature will have leveled the Billerica dam, and the Lowell factories, and the Grass-ground River [the Indian description for the Concord River] will run clear again, to be explored by new migratory shoals, even as far as the Hopkinton pond and the Westborough swamp.[12]

10

GRID ON THE MOON

"You know, you discovered Khuzistan for us Iranians. We didn't know about it. You are the Columbus of Iran, just as another Columbus discovered your country. Your country was there all the time but Columbus discovered it. Khuzistan was there all the time, but you found it," said the prime minister of Iran to David Lilienthal.[1] Lilienthal knew this was rank flattery, but he appreciated it anyway.

Working in the heart of darkest Tennessee in the 1930s, Lilienthal had helped make a poor and sparsely electrified region blossom. The balance of his career, after he left U.S. government service in 1950 and for the next 30 years, was essentially devoted to trying to re-create the Tennessee Valley Authority (TVA) experience in other countries. He was a self-appointed do-gooder. In Khuzistan province, a dry, resource-rich region in the west of Iran, a place where King Darius had built a network of canals, a sort of water grid in the 6th century BCE, Lilienthal helped inaugurate a major development project, including the Dez Dam, 600 feet high and generator of 500 megawatts of electricity.[2]

Lilienthal was much in demand as adviser and manager. On the same 1969 trip that took him to Iran he flew to the Ivory Coast, where he conferred with the country's president. The agenda concerned, as it usually did, construction of a dam and development of an underused river valley. Lilienthal's urged the early relocation of people living in the area soon to be turned into a lake behind the dam. Otherwise, Lilienthal said, when villagers saw the water rising, they might panic. Start moving them now; don't overplan; don't dither over details. Don't just supply subsidies, because these tended to become permanent; reassure the villagers about the benefits of electricity. "Emphasize the human and emotional side, not the scientific," he said.[3]

A pattern recurred wherever Lilienthal went. He would be warned by outside observers about the fragility of a poor nation's government, its corruption, its disregard for its own citizens. But Lilienthal would go anyway. He would be flattered by the officials and then tender his advice, which often was ignored or stymied. In country after country, some progress was made, some people would receive electricity, but millions of others did not.

MINIMUM ELECTRICITY

Life is unfair. Not everyone who wants electricity can get it. The grid, in my extended sense of the word, has referred to both the electrical network itself—the wires and the generators—and the electrical culture that goes along with it. The grid today might be worldwide, but it's not universal. Why that is, and where the grid will go next, will be the subjects of this chapter.

Is the grid a luxury? Even as this history of the grid begins to wrap up, there are still plenty of questions to address. Is electricity an absolute necessity of life? Is it a human right? The answer to both these questions is no. Great civilizations of the past obviously got on without electricity.

No, electricity is not an absolute requisite for life, but it helps a lot. Having more electricity correlates with higher literacy, longer life expectancy, better nutritional intake, and lower infant mortality. Furthermore, a peasant's lack of electricity doesn't necessarily mean that he makes more sparing use of energy. In fact, it was often the case that preindustrial dwellings used more primary energy per capita than their 20th-century counterparts.

How can that be? Because people used what was at hand, often wood, a meager, polluting form of fuel. Indeed, most primitive forms of energy use were for heating, cooking, lighting, pumping, or traveling. Even now, the amount of primary energy required to produce a unit of gross domestic product is typically three times *higher* in poor countries than in rich countries.[4] The difference between old and new is particularly stark for lighting. Electricity made a 10-fold and then a 100-fold improvement in the amount of illumination per energy used over methods using flames.

This should settle the issue of whether electricity is a luxury. Having it saves energy and improves lives. Not having it hurts. True, you can have too much electricity, or at least make wasteful use of the electricity you have, but how about too little? What is the *least* amount of grid you need for a decent life, decent meaning being able to eat, sleep, and read in a relatively safe place of shelter? The least you need is 1,000 kilowatt-hours per year per person, says the Electric Power Research Institute, a nonprofit organization founded by the U.S. power industry and based in Palo Alto, California. In its comprehensive report, *Electricity Technology Roadmap*, EPRI argues that for basic lighting, communications, refrigeration, and local agriculture, you need 1,000 units of electricity—call it minimum electricity—to *begin* the journey out of poverty.[5]

What is a kilowatt-hour? It means you're using 1,000 watts of power for an hour. You could burn ten 100-watt lightbulbs or run a large air conditioner for an hour. You can wash 200 pounds of laundry or mill 300 pounds of grain. It takes about 1 pound of coal to produce 1 kilowatt-hour.

Having 1,000 of these power units per person is a threshold. Having this amount of electricity doesn't guarantee a good life; it's the amount you need to *get started* toward a good life. According to EPRI, roughly 4 billion out of the 6 billion people alive today get less than minimum electricity. More than a billion have no electricity at all. The *Roadmap* goal is for everyone on earth to have minimum electricity by the year 2050.

A thousand units of electricity is, by the way, just about the amount the average person had in Chicago in the year 1925. In other words, the minimum electricity level for everyone on earth, by no means an easy thing to achieve even by the target date of 2050, would in effect bring the poor African village of today no further along in technological terms than Chicago in the Roaring Twenties. By comparison, the typical American's intake of electrical energy right now comes to around 13,000 kilowatt-hours per year.

Thus far we have looked mainly at the one-third of humanity that consumes electricity at better than the minimum rate, so it is appropriate for a book about the electrical grid to look also at the far-flung two-thirds of humanity that fall below the line. By far-flung I don't mean in the sense of being geographically or culturally distant from the Europe–North America–Pacific Rim centers of economic ascendancy, but in the sense of being distant from the grid. Like a person excluded from a warm place on a cold day, if you are far from the advantages of electricity, you can be said, figuratively, to have been left behind in the year 1925 or the Middle Ages or earlier. You have been held back while the rest of the class moved on.

In the early part of this grid narrative, when Thomas Edison filled wires with energy beneath Manhattan's streets, this was the morning of the grid, so to speak. But for a village in India getting its first electrical hookup only now, it is *still* the morning of the grid. For millions of others, those not scheduled for power installation over the foreseeable future, it isn't even morning yet but rather somewhere before the dawn.

SMELTING IN UGANDA

Some of David Lilienthal's projects succeeded, but many others foundered or performed disappointingly because he met up with some potent real-world forces in his client countries: entrenched commercial and bureaucratic interests, mediocre local management, envy and resentment of his access to senior officials in the country's government, insufficient funding, and exaggerated expectations. All these forces had also been at work in the struggle to create and maintain the original TVA, the great model so admired around the world. So why hadn't it worked elsewhere? Perhaps Lilienthal's earlier success could be attributed to conditions unique to that period in American history or to having a vigorous patron in the form of Franklin Roosevelt or to the later wartime need for TVA's plentiful electricity. An additional explanation of why TVA was not reproduced abroad (nor in the United States itself), was that the world was getting ever more complicated. The Cold War, the end of colonialism, racial sensitivities, chronic poverty, high illiteracy, and epidemics were factors in tarrying the course of technological development.

Lilienthal thought that if he could arrive with the cavalry—in the form of his team of experts, loans from the World Bank, and the beneficence of the local leader—he would be able to overcome decades of stagnation and make the deserts bloom. Unfortunately, it usually didn't work that way. Tennessee got electricity, but many other places did not. Many poor places stayed poor.

One of the poorest of the poor places is the so-called Great Lakes region in East Africa, the cluster of countries—Burundi, Kenya, Rwanda, Tanzania, and Uganda—surrounding Lake Victoria and Lake Tanganyika. In Uganda the first notable grid development came in the 1950s with the advent of the Owens Falls Dam, a construction effort so big that it took most of nation's existing electricity just to get the job done. Even when this dam was finished, there wasn't much of a grid. Most of the power went to a few big industrial users, such as the Kilembe copper smelter, the Nyanza textile factory, and the Sukulu fertilizer plant.

Compare 1950s Uganda with 1880s New York. Edison's accomplishment was to replace small kerosene electrical generators with a unified, efficient central grid. That was also the aim for Owen Falls. New York in the 1880s was wealthy, whereas Uganda in the 1950s was poor. There were plenty of rich customers in New York who could afford the high rates, and there were millionaire financiers like J. P. Morgan who could put up the money for the vast infrastructure needed. In Uganda there is no Morgan and there were few wealthy customers.

Furthermore, when it comes to electricity in Uganda, race is a factor. Europeans in the country (during colonial times) had most of the energy. In Western electricity terms, Europeans in Uganda in 1950 lived at the 1920s Chicago level. Asian residents had the next most electricity; for them it might have been the year 1895. Coming in a distant third, black Africans generally had less than minimum electricity. You couldn't even specify an equivalent year.

In a 1950s study of expected power patterns for the coming years, the Uganda Electricity Board began by saying that most European residents already were connected and that by 1970 the Asian residents would catch up with the Europeans. What about native Africans?

> The actual benefit which the African may derive from a given commodity may be very much greater to him than to a European, because it represents such a very great advance on what was previously available, or because of the circumstances under which he lives. The bicycle, for example, has revolutionized in no small way the manner in which the African peasant lives. The tin roof, which gives protection; the radio, which widens his horizon; and the electric lamp, which transforms the interior of his hut—all of these things have made such a profound change to the African's life that the value which he puts upon them is relatively very much greater than with the European.[6]

The African is different, the report intimates, because he is starting the great footrace of life further behind. In other respects, though, he is just like other people: "The African's liking for consumer goods such as radios, electric lamps, irons, appears to be quite as strong as the Europeans'." The report went on to predict that the power from Owens Falls would be fully subscribed by 1965, by which time more dams would be needed. The prediction proved correct. So how do Ugandans get electricity nowadays? They don't. There is no coal in Uganda. The primary energy source in Uganda, besides hydropower, remains wood and animal waste. The available electricity still goes largely to support industry. Under these circumstances, how are people in Uganda supposed to achieve minimum electricity? No one knows.

Massoud Amin, one of the authors of the Electric Power Research Institute report, suspects the answer will be a combination of old and new. Old—high-voltage power brought in by large transmission lines from the north end of the continent (rich in oil energy) or from the south end (rich in hydroelectric and nuclear generation) toward the energy-poor midsection of Africa. New—with better and cheaper renewable-energy machines, like wind turbines and solar panels or solar concentrators; Africans might be able to bypass the older, capital-intense, wire-heavy, fossil-burning habits of the grid elsewhere, just as in some places cellular telephones are coming to places that never had wired-up landlines to begin with.

Amin knows from experience growing up in Iran how the installation of electricity could, in just one generation, boost a village from Medieval times into the 20th century. This boost has not yet happened in Uganda, where even now only 3 to 5 percent of people have regular access to the grid. An enhancement of the Owen Falls Dam (now called the Nalubaale Dam) came into operation in the year 2000, but further dam projects supported by the World Bank are presently on hold while allegations of corruption are investigated.[7]

For electricity in Uganda, in the Lake District of East Africa, it is still the morning of the grid and will be for some time to come.

SITUATIONAL AWARENESS IN OHIO

Massoud Amin spends much of his time, as a professor of electrical engineering at the University of Minnesota, thinking about large electrical networks, the fat part of the worldwide grid where customers use electricity in amounts far above minimum electricity. He looks at the grid like a doctor examining a patient, who can be malnourished by eating too much or too little. The African patient is underweight, while the North American patient tends to be overweight. The North American consumes twice what a European needs for essentially the same quality of life. Too little electricity and economic performance is anemic. Too much electricity brings about different problems.

In Ohio there is too much electricity, or at least on some days the customers in Ohio want more electricity than is good for the existing grid. The grid in Ohio is more advanced than the grid in Uganda, but they have some things in common. Ohio is not tropical, but it does get hot in the summer, and the heat encourages people to turn on air conditioners like mad, which means more flowing current, which causes transmission lines to heat up and sag as the metal wire expands. Ohio does not have Uganda's jungle, but it does have plenty of untrimmed trees which, reaching up, can make unfortunate contact with the electrical grid drooping down. Uganda does not have many of Ohio's advanced computers, but if the computers you have are ill used, Ohio may as well not have them.

That was exactly the situation on August 14, 2003. A few tree-limb short circuits, coupled with computer errors and plenty of human operator negligence, led to a critical loss of "situational awareness." In other words, the grid operators didn't know what the hell was happening. Some of the lines stopped functioning. Other lines became overburdened. The electrical instability then spread from one grid jurisdiction to another. Suddenly the gigantic electrical failure featured at the beginning of this book was born. It was the failure of November 1965 all over again. Too much power in the lines, then too little.

In 1965 the dominoes fell one after the other—Buffalo, Rochester, Syracuse—the territory of the old Iroquois nation. In 2003 these same cities, and many more besides, were participants. Some of the same power lines were involved. The very same electrons would have been yanked around on those two occasions separated by 40 years, the first and second acts of what could be considered a single slowly-unfolding electrical drama.[8]

CARNIVALS IN ROME

As bad as the electrical collapse was in Ohio, somebody somewhere always has it worse. Uganda is worse. In August 2003, Detroit's or Cleveland's grid made a full or nearly full recovery within a day or two. In Baghdad, where the fighting that followed the American invasion of 2003 obliterated much of the oil and electricity systems, things were far worse. Asked what he thought of the big power failure in the United States on August 14, one Iraqi resident said that mostly Americans live like kings. If they are bothered by a few hours at temperatures of 90 degrees Fahrenheit without air conditioning, he said, they should try living for months at 110 degrees.[9]

Europeans were especially incredulous as to how the North Americans could have such frequent and extended blackouts. Yes, in the month or so after the August 2003 blackout there were minor power disturbances here and there. A million lost power in London for a few hours, 4 million in Denmark and Sweden. But the general feeling was that an avalanche on the order of the North American disasters—30 million people cut off in 1965 and then 50 million in 2003—could never happen in Europe.

Can the failures get bigger? Sure they can. Complexity science suggests the *how* but not exactly the *when.* On September 28, 2003, hardly six weeks after the August 14 extravaganza, almost the entire Italian grid collapsed. An estimated 57 million were without power. The blackout struck at 3 a.m. during an all-night culture carnival in Rome called White Night. Thousands of people were forced out of museums or artistic performances into a heavy rainfall. How had it happened? They weren't sure at first, but it seemed to involve French-produced power passing through Switzerland and then on into Italy. At one point in the public relations battle, the Italian grid operator, GRTN, was blaming France. The French company, RTE, blamed the Swiss company, ATEL, while the Swiss blamed the Italian grid operators.[10]

Despite the Italian event, Europe has had a better overall record in avoiding large blackouts than North America, owing not least to the larger margin of reserve electricity (standby generation) in Europe. These extra reserves have a price, however. You get what you pay for. Large blackouts could be mitigated, if not exactly eliminated, by paying more for better equipment and training. The sensors that watch for trouble signs could be made much faster and better coordinated.

Massoud Amin has worked many years on designing such a smart grid. Implementing an alert system, one using fast-responding, Global Positioning System synchronized monitors and adroit current-switching equipment would be expensive, costing something like $10 billion per year for 10 years in the United States. But the cost of *not* making the upgrade is far higher. Estimates of the damage report for electrical disturbances to the U.S. economy range up to $100 billion per year.[11] A majority of this loss comes not from the headline-grabbing regional disturbances but from smaller outages or voltage irregularities lasting minutes or seconds. Even a fraction of a second is enough for an electricity glitch to turn off a computer, leading to a business loss of millions of dollars for companies like microchip manufacturers or processing centers for credit card sales.

Keep in mind that running the grid at an efficiency of 99.9

percent—a level referred to as "three nines"—sounds pretty good for any human-built system, but this still leaves nine hours of downtime per year. That can be a little or a lot, depending on who you are or what you do for a living. We get the grid we deserve. Blackouts can and will get larger. They already have.

REFRIGERATORS IN BOMBAY

Ugandans would love 99.9 percent electricity. They would gladly take on Ohio's grid problems. In Ohio everyone has the grid, in Uganda almost no one. In Ohio it might be said that some of the computer control equipment used to operate the grid is 15 years out of date. In Uganda large parts of the grid are 90 years out of date.

Somewhere between Uganda and Ohio is India. Is electricity in India different from Western electricity in the way that the Indian elephants are different from African elephants? Is power in India, the subcontinent beneath Asia, different from power in Italy, that peninsula hanging south from Europe? No, of course not. Electricity is the same everywhere. You spin a turbine, induce a high-voltage difference across a set of wires, and send power out among the citizens, who use it to warm filaments and activate machines. The voltages might be different and the arrival rates might be 50 or 60 waves per second, but otherwise there is no difference from place to place. There is no Hindu electricity, or South American electricity, or electricity just for women or kings.

Yet with so many people below the minimum electricity line, with millions having no access to the grid at all, the aggregate electrical experience *will* be different from Ohio. In India, as in Uganda, it has been the morning of the grid for many years. Here's the way it was in the 1930s:

> I am often asked what are the greatest changes which have taken place since I landed in Bombay in July 1897. Undoubtedly the most beneficial is the advent of electricity . . . nothing has increased so materially the amenities of life in the City or Island, and raised the standards of health more, than the general use of electricity.

Thus begins a history of the Bombay Electric Supply & Tramway Company (BEST) prepared on its 50th anniversary in 1936. Despite a tone that indicates a very slight disdain for the "softness" in living brought about by the grid, these introductory remarks, reveal an evident pride in the spread of electricity not just to affluent customers but to the common citizens:

> It is not generally realized that through the very low rates it charges to small consumers—I might even say unprofitable rates—the B.E.S.T. has carried the benefits of electricity down to a humbler class of citizen than any other corporation in like circumstances. There are many thousands of users of electricity whose monthly bills are under two rupees, and statistics show that a much higher proportion of the population is connected with the supply system, as compared with any other city in India. Last, but by no means least, we have the electric refrigerator.[12]

Having the refrigerator was a much more advanced stage of electrical ownership than mere bulbs, since coldness in summer is more difficult to achieve (because it requires moving parts) than illumination after nightfall. It made you a stakeholder in the new order. It meant that a perishable bit of food might thrive for days. A chilled drink could set you apart from those in the home next door. Unfortunately, many Indians did *not* get a refrigerator and have lived, in an energy sense, as if it were still Bombay (now Mumbai) in the 1930s or earlier.

Bihar state is particularly poor and far from being uniformly hooked up. Even by 1980 the fraction of villages served had only reached 30 percent. In other states the numbers weren't much better. The main problem is poverty. The states don't have the resources to wire the villages, much less send electricity on a regular basis. Customers would not have been able to pay at the prescribed rate anyway, and indeed tribal life was not at all arranged around the presence of electrical energy. You'd get up at daybreak, work as long as there was light, and then go to sleep. When electricity first came to a village, it would be cause for celebration. People were drawn to a home to see the first lighting of the bulb. It was said that a god had descended on the house.

The first benefits of electricity in small-town India were pretty much the same as in small towns on other continents. Electric lights extended the work day, making after-dark dinners possible. As in other electrified parts of the world, saying goodbye to smelly, expensive kerosene was a pleasure unless, perchance, the new electrical setup proved unfaithful. In such cases, some said, it was better not to have electricity at all. Sometimes the service was offered but refused. Among the reasons given for *not* taking service: "communication gap between people and functionaries . . . fear of exploitation . . . afraid of shock and afraid of catching fire."[13]

Utility customers the world over have at various times complained of slow restoration of service after a power failure. Twice my own neighborhood has gone without power for a week following storms, a situation that would not be tolerated in Europe. But then there is the old refrain—things can be far worse. Consider the case of several villages in West Bengal. After a storm, some power poles were knocked down and regular electricity vanished. In the first weeks and months, the villagers complained and were promised a fix. More than 22 years later, they were still seeking a restoration of service.[14]

Many in India are far-flung. Hundreds of millions are far from the grid or only weakly attached. On the other hand, it can be said that hundreds of millions *are* plugged in, and this makes India a major electrical nation. It is the third-largest producer of coal and the sixth-largest energy consumer in the world.[15] But does India resemble the robust-economy countries in having a complex multiconnected grid vulnerable to cascading failure?

It does. On January 2, 2001, an event exceeding the Ohio or Italian collapses arrived. If superstitious you would say this was a bad omen coming only two days into the new millennium by the official calendrical reckoning. The failure of a transmission line at a substation in Uttar Pradesh, the most densely populated state, brought down a considerable portion of the grid across northern India. The number of people left without power was estimated to be 220 million.[16] This bears repeating: Because of a malfunctioning piece of electrical equipment, the largest shutdown of electricity in history, in terms of people deprived of power all at one go, came into the record book. Nearly a quarter *billion* were affected. (Possibly a larger number of people were involved in an earlier blackout in West Bengal in June 1990—but numbers were hard to determine.[17])

An estimated 15,000 megawatts was shut off during the 2001 event. Compared to the U.S.–Canadian outage of 2003, the Indian blackout stranded roughly a factor of four or five times *more* people but a factor of four or five times *less* power. In other words, electricity is spread more thinly on the ground in India than in North America.

There are several reasons for this thinness. One is the remaining persistent high number of rural residents without electricity. Another is the near bankruptcy of the state power companies, which, to satisfy popular demand, keep rates artificially low. Because of this less-than-breakeven revenue stream and a rate of pilferage (illegally tapping of the lines for power) in some places as high as 20 percent, the state-run regional grids are in a weak position. In Uttar Pradesh, where the huge outage took place, no new electric generation had been added for the preceding decade, although many new mouths to feed had come along during the same period. Consequently, outages are common: 10 to 15 minutes every other day. For savvy companies that can afford it, expensive standby diesel generators are the solution.[18] For those that can't, there is darkness.

Like all other technical networks, the Indian grid is built on the side of a sandpile, as it were. A small disturbance in the network can lead to a major disturbance in delivery. The financial footing of the Indian grid is also very sandy. Consider the difficulties arising from the need for a poor company in a developing country to seek private investments from abroad. The lender naturally seeks a good return on its investment and guarantees to protect the loan in case of default. The borrower naturally seeks the easiest terms and loosest guarantees.

This brings us to a painful case history. A consortium of lenders, led by the U.S. company Enron, signed a contract with the electricity board in Maharashtra state to produce an immense power plant fired with liquid natural gas. The new project, under the auspices of the Dabhol Power Corporation, ambitiously hoped to produce as much as 2,400 megawatts of power, as much as the Ravenswood plant in New York, home to Big Allis.

Alas, the multibillion-dollar deal, the largest foreign investment project at the time in India, was troublesome from the outset. Complaints were made that the contract was signed too hastily, that the plant charged too much for power, that it produced electricity whether it was needed or not, and that its construction violated various environmental and safety standards. Then the state electricity board fell behind in its monthly payments, and legal proceedings began. At the time of the gigantic January 2001 blackout in Uttar Pradesh, the plant was asked to ship power northward to help with restarting the grid, but Enron's power plant, still seeking to be paid for previous service, insisted on an emergency rate three times higher than normal.[19]

As India sees things, foreign firms seek quick and dirty profits. As foreigners see things, India is an unreliable place to invest. Result—companies such as Electricité de France have cancelled or pulled back on plans for power investments in India owing to worries about payments or the very solvency of the state electricity boards. (Enron would, in a matter of months, be having solvency problems of its own.)

Making electricity and selling it at a price the customer can afford and that allows the company to expand has been a struggle for utilities in all lands ever since the first grids went up in the 1880s. More than a century later it's a special struggle in the subcontinent. How India can produce power for a burgeoning population, attract the involvement of outsiders and their much needed expertise and cash, and yet remain sovereign is a problem still to be worked out. In countries with a mature grid, an important issue seems to be improving reliability from 99.9 percent to something even better. In countries like India the chief issue is that of mustering resources for a nation where large sections of society do not have minimum electricity or where service regularity is not at the level of three "nines" or even two. In many places it is at the zero level.

HYDRAULICS IN CHINA

China can't be left out. India is immense, but China is vaster still. No account of electricity or technology or energy procurement or the heavy toll expected from such a command of resources can fail to include China. Its economy has bounded forward over the past decade, growing in some of those years at double or triple the rates of most Western nations, with electrical demand growing 10 percent or even 15 percent every year. Long a poor country, and still a poor country in many respects, China wants to catch up with the industrial world, and it is well on its way.

China sprawls out in time and space. It has all the zones you can find in a geography textbook: mountains to the south and north, deserts to the west, steppe and grasslands to the north, and broad floodplains to the east and south. The Chinese spectrum of electrical infrastructure extends from 1900-level diesel-generated electricity in hamlets to 1940s-era coal-fired heat-and-power production for keeping 24-hours-a-day steel mills rolling out their product. At the top end they're building late-model nuclear reactors and wind turbines for supplying power to the bustling eastern coastal metropolises now competing with Manhattan for skyline honors. More surprising still is the fact that high-growth's usual sour companion—monetary inflation—has remained low. What then is the price for sudden success?

There is always a price. In China now, for example, highways are much more congested, partly because there are more people in the cities and partly because millions are trading in their bikes for cars, repeating the worst excesses of American consumerism. Of the 10 most air-polluted cities in the world, seven are in China. With this atmospheric burden comes a related increase in respiratory disease—300,000 air-pollutions deaths a year.[20] If the worldwide embrace of gas-powered electric plants (with their lower CO_2 emissions) over the past decade or two represented a "decarbonization" of sorts, then the prospective escalation of conventional coal-powered plants in China and India (and Idaho) represents a grim sort of "recarbonization."

It's tempting to look past the downside and concentrate on the positive. Look at what is happening to the average Chinese urban woman. She now possesses more personal floor space than before; uses more water for drinking, washing, and other chores; eats more meat (increasingly beef); owns more machines and requires more electricity to power them. She wants her own automobile and generally aims to get what the modern woman in Italy or America has. China is the world's largest producer of electrical gadgets.[21]

As measured by actuaries, living is improving in China. Life expectancy has overtaken that of Russia. Gross domestic product is equivalent to that of 1950s Japan. Infant mortality has dropped to the level of Argentina. China may well arrive at the recommended per-capita electric consumption of 1,000 units well before the target date of 2050. India maybe. Uganda probably not.

The toll for China's turbo expansion is heavy indeed. It is now the largest user of chemical fertilizers, which are good for vegetables in the cultivated field but bad when the fertilizers run off into the rivers and feed blooms of algae. China is the world's largest excavator and combustor of coal and perforce the greatest releaser of sulfur compounds, which come back to earth in showers of acid rain, helping kill forests and deface buildings. The nation is being drained, shrunk, parched, leached, nibbled, silted up, fouled, and turned to desert.

With enough engineering and scientific innovation thrown into the national effort, and with enlightened political management, maybe environmental depredation can be contained, allowing for sustainable growth. But not if you have no trees or topsoil. China's per-capita cropland is half the world average, not much different from the low level of available land that helped trigger the genocidal struggle in Rwanda. China's cup is no longer even half full. Its per-capita fresh water is one-fourth the world average and its per-capita forested area only one-fifth.[22]

Too much human energy and not enough electrical energy. High expectations and ambitious industrial plans but insufficient resources. And even if the electrical generators can be mustered and coal pulled from the ground more quickly and shoveled into the furnaces, there is still the problem of how to suppress the billows of pollution coming from those generators, which are gradually tinting the air above the new eastern skyscrapers an unpleasant shade of brown.

Perhaps the most conspicuous response to the combined problems of electric shortage and pollution surplus is the dam being built on the middle reaches of the Yangtze River. Called the Three Gorges Dam after its picturesque setting, the project is the largest electrical construction job ever undertaken. Here the big numbers overtake all the other big numbers used so far to describe China's big push forward. Two kilometers wide, 175 meters high, costing a minimum of $30 billion, the dam is a gigantic experiment in flood control and power production. The 26 turbines, when they're all in place, will pour 18 billion watts into the high-voltage grid. This is half again more than the output from the world's largest existing hydroelectric plant, the Itaipu Dam on the Parana River near the Brazil–Paraguay border, and three times the output of the Grand Coulee Dam, the largest hydroelectric installation in the United States.

Already the waters are rising in the giant reservoir lake behind Three Gorges, and the first turbines have begun to operate. A building project as big as Three Gorges, often compared in magnitude (and maybe in folly) to the great pyramids, is a provoker of opinions and fantasies, even within China, where information flow can be severely controlled. Some critics fear that this largest of generators will lead to massive avalanches on the Chinese grid. Could we see the first billion-person blackout? Three Gorges's pollution-free electricity will be highly prized by economists who plot the fortunes of factories and the schedules of trollies, but naturalists foresee a great silting of waterways and a wholesale loss of habitats and farmland. Demographers, and certainly the residents in the way of the rising waters, have taken note. An expected million-plus migration is now taking place.

In China there is great pressure to keep enlarging the economy, although maybe not quite at the recent red-hot level, so that the Chinese people can enjoy electrical levels comparable to those of other leading nations. And why not have these aspirations? As for the extra pollution brought to the cities by the new cars, the new factories, and new electrical generators, well, this problem can also be put off to another day. It's tempting to pay for today's needs by borrowing against tomorrow.

Such a delay would only be part of a larger electrical leadership dilemma. In Ohio (with pollution of its own) costly transmission line upgrades and a sharpening of situational awareness need to be made, but political realities dictate that these not be started until "next year" at the earliest. In India they need to raise electric rates to the breakeven level and attract foreign investment. Maybe next year. In Uganda, which comes first, good government or good dams? Suspicion of corruption holds up construction of a new hydroelectric plant that would greatly enlarge the national grid, and along with this national literacy, and along with this democratic institutions. When will this happen? Perhaps starting as soon as a year from now—or maybe not.[23]

Massoud Amin, besides his other duties and research interests, is director of the Center for the Development of Technological Leadership at the University of Minnesota. He says that when his students—who continue to hold down jobs in engineering or construction companies during the leadership training—consider which nation should be picked as the site for their foreign-visit instruction, they invariably propose China: big nation, big economy, big problems, big opportunities, big potential for practical leadership lessons.

How big is China? How important is it? To illustrate the bigness, I offer a personal anecdote from not so long ago. The occasion was an Olympic diving event being shown on television. The man favored to win, an American, had come in first in the previous competition. This time around he was being pushed to the limit of his ability by two Chinese athletes. China, viewers were told, had possessed an Olympic diving team for only a few years. And yet the Chinese team was suddenly close to winning a gold medal. It was a dramatic catching up. As it happened, the American diver did win the contest; a Chinese diver came in second. China had won respect and something else, since it had succeeded in changing the existing order of things. This time it was springboard diving, but soon perhaps it would be other endeavors—designing computers or inventing a nonpolluting way to make electricity.

If we can operate on the assumption that innate talent is spread evenly around the world—talent for creating art, say, or talent for springboard diving, or for engineering—then the nations with the largest numbers of people should, if they choose to apply themselves in that particular area, enjoy the greatest success. It would seem that if not now then eventually, all other things being equal, China would come in first and India second. The medal would not be a lump of bronze worn around the neck but would be prosperity, maybe even economic hegemony over other nations.

Ah, you say, but things are not equal, and national cultural differences, and economic opportunities, are still quite varied around the globe. Chances in life are different for different places. Yes, but not as different as before. Internet technology, relatively cheap aviation, and growing educational flexibility have made the world smaller and larger at the same time.

So here we are, riding the present moment through life as we always do, poised between past and future. The Chinese diver stands at the edge of the springboard preparing to take his turn in the competition. And behind him, mounting the ladder, are other people from other lands who wish also to move up in the standings.

SYMBOLIC ACTS

China catching up to Western levels of electrification—this might be a logical place to end the story. Or how about finishing with a quick summary of the latest inventions, the proposed technical fixes for all those billion-dollar blackout problems? Well, this is not how the book will end. I am not going to profile some promising new thermo-electric turbine design since, anyway, it is likely to be overtaken soon by yet another model. Nor am I going to specify how and when minimum electricity goals will be met in poor countries. Nor speculate on when or how the American Congress will seriously address the grave connection between energy use and climate change.

Instead the book will end with a search for more meaning. What does the grid *mean* for the human race? We haven't exactly avoided this question. We've looked at some of the best things the electrical grid does. It helps yield more food per acre, it helps scan for tumors, it makes possible cheap aluminum and kerosene-less lamps for reading after dark. We have also had occasion to look, sometimes with the help of Lewis Mumford, at the social compromises brought on by electrical equipment (with much help from other fossil-fuel-burning technologies, such as automotives): not quite clear skies, no more Milky Way, and less self-made musical entertainment. No more piano.

When will we know if we got the right mixture of benefit and compromise? Does the grid serve any purpose other than bringing convenient energy by wire? We can't fully answer, but we can begin to consider these rhetorical questions by pondering yet one more grid, perhaps the most profound and important electrical grid of all—the one inside our skulls.

The functioning brain, assisted by the rest of the central and peripheral nervous systems, constitutes a formidable electrical network.

Like the utility grid, the neural grid has wires, an estimated 100 billion neuron pathways—as many cells as there are stars in the Milky Way. Moreover, each neuron can form as many as a thousand linkages to other neurons. Like the utility grid, the neural grid carries enabling waves of electricity, in the form of electrochemical pulses. The speed of these pulses, up to about 300 feet per second, isn't close to the speed of light but is fast enough to regulate the moment-by-moment functions of the body: breathing, seeing, heartbeat, sense of balance, and so forth.

An even more profound use of the bioelectrical grid is to energize the mental state we call consciousness. Without the neuro grid we wouldn't have the utility grid or the telephone grid or the Internet or any other electrified network. All these grids are *made* things. They come late in the long stretch of historic and prehistoric time over which men and women made many things: locomotives and steam engines, and before that aqueducts and looms, and before that pyramids and ovens; and before that clothing and bread and the wheel, and before that spears and stone hatchets.

You could say that man made tools and that, in a certain sense, tools made man. According to this line of reasoning, one large factor in the difference between humans and other animals is the much greater use humans make of tools. The very art of making and using tools might have spurred intellectual development over evolutionary time, leading to better tools, encouraging further mental development, and so on in a positive feedback loop of advancement. First the wheel, later high voltage.

Is the electrical grid just one more of humanity's most recent and more elaborate tools? Did the early use of tools make a woman or a man more human, more intelligent? Are tools the key? Lewis Mumford thought not. In his extensive meditation on human history he decided that what mattered even more than the propensity to make tools was the human tendency to fashion symbolic reality, to devise games and rituals and myths and, most important of all, language.

Before manual dexterity, Mumford, says, came mental dexterity. Before man the maker (Latin: *homo faber*) came playful man (*homo ludens*). The greatest human accomplishment was not the development of agriculture or the discovery of fire or the use of stone axes. It was the very act of becoming human, of coming into a state of self-regarding consciousness, a supreme process that took millions of years:

> The technological feat of escaping from the field of gravitation is trivial compared to man's escape from the brute unconsciousness of matter and the closed cycle of life.
>
> . . . The humblest human mind encompasses and transfigures more conscious experience in a single day than our entire solar system embraced in its first three billion years, before life appeared.[24]

The story of the electrical grid is part of the story of technology, which in turn is part of the larger epic of human development. It's this larger cultural realm—much larger than Ohio or Uganda—that I should like to examine here at the end. Concluding the history of the grid is problematical since it is like aiming at a moving target. Therefore, I won't conclude; I'll continue toward a kind of perpetual finishing. What I propose to do is to move the story backward in time, to the year 1969, and then come forward again as if we were moving into the future. In this way I hope to sidestep Kierkegaard's aphorism about life being lived forward but understood backward.

In this finishing, if not concluding, section of the book we are going to look at one final grid-related event, an occurrence that represents the most observed, the most psychologically intense, and, to my way of thinking, the most *meaningful* use of electricity in history. To do this we'll have to go to the moon.

THE GRID IN 1969

The more prominent explanations for rocketing men to the lunar surface were the following: (1) *Exploration.* It is in man's nature to explore, and the moon's surface was the grandest destination within the technical means of current civilization. (2) *Military.* The moon expedition was an extension of America's hot war being waged against the North Vietnamese and the Cold War against the Soviet Union. (3) *Prestige.* It underscored America's resolve to fulfill huge undertakings and its ability to be the first to do something so difficult. (4) *Big science.* The moon could be a place where scientific discoveries might be made and valuable mineral deposits located. (5) *Big business.* The military–industrial complex of high-tech companies, not content with lucrative war work, desired still more contracts for their eager factories.

Bringing electricity to the moon was *not* one of the stated reasons for the sequence of space flights known as the Apollo program. Nevertheless, electricity was a major subsystem of the spacecraft itself, as indeed it must be for any substantial modern device of many parts, or one requiring subtle maneuvers at high velocity. A comparison between Apollo and that other epitome of 1960s advanced engineering, Con Edison's Big Allis dynamo, will illustrate the point. First, they're both about 400 feet tall, Allis with its stack and Apollo on its pad. They both burn immense amounts of fuel, Allis to make electricity and Apollo to reach the moon. One takes for granted that Allis is filled with metal conductors. So is Apollo, with 15 miles of electrical wiring.[25]

Both Allis and Apollo are complexity machines, small mountains of components linked to other components, many of which take part in sequences of actions leading to other actions. This doesn't mean that Allis and Apollo are necessarily badly designed or that an accident is destined to happen, but only that the tightly coupled complexity inherent in their architecture gives them some of the tendencies of a sandpile in which avalanches *will* occur. Even a tiny slipup, a tiny perturbation to the system, could lead to a large failure.

Such it was with the first Apollo mission scheduled for flight. The perturbation that set the avalanche in motion was slight. It was an electrical shortcoming; the sequence of events was quick; and the results were catastrophic. Lying in their takeoff prone position in the main capsule, three astronauts were being fed pure oxygen, the better to keep cabin pressure low while supplying the men with the needed breathable substance. Somewhere else, in an equipment bay in the capsule, an electrical short (electricity trying to take a shortcut) is believed to have sent sparks onto some combustible material. A fire began, and in the rich oxygen environment the flames raced around. After all, you brighten a fire by blowing on it, by lending it more oxygen. The men were burned alive.

The families, the corps of astronauts, and the nation were stunned. There had been fatalities before among American astronauts and Russian cosmonauts. Whether test piloting new vehicles or riding a rocket into space, the men (and later women) faced many hazards. Still, the deaths of those three men was the biggest blow yet to the manned spaceflight program. NASA's troubles could be seen as part of a culturewide unease. There was at this time, the late 1960s, a vocal protest movement in America centered around disenchantment with the Vietnam War and the campaign for civil rights.

Many people, not just protestors, now regularly questioned whether billions of dollars should be spent on a new Lewis and Clark expedition to the moon when Detroit and Newark were ablaze with poverty and racial tension. If you suggested to residents of those cities that a contraption with men inside was being dispatched to the moon and that because of this far-flung beachhead the collective consciousness of the human race was about to be enlarged, they might have thought it was a grim joke.

The moon mission went ahead anyway. Apollo's wiring was redesigned, the oxygen was replaced with something more like normal air, and the materials inside the crew compartment were made fireproof. A later spaceshot, *Apollo 8*, helped rekindle enthusiasm for NASA's exploration extravaganza. On this flight, humans had for the first time ventured all the way to the moon. Although they did not descend to the surface, they did record those famous pictures of the earth rising above the limb of the moon. An astronaut could and did hold up his arm and with his outstretched thumbnail cover the entire view of his home planet and along with it all the present inhabitants thereof, and all the oceans and landmasses, and all the strife and the achievements.

While some earthlings were preparing to stroll across the dusty plains of the Moon, back on *terra firma* the subways beneath the streets of New York still had to run. Elevators, stoplights, and network television broadcasts all required regular electricity. Consolidated Edison, the company with the monopoly right to sell power in the city, had the statutory duty to fill all requests for voltage, large or small, day or night. Still on the rebound from the 1965 blackout, constrained by new environmental regulations, and harried by citizens associations against locating any new generators in *their* neighborhoods, Con Ed was up against it.

Con Ed, always thinking ahead, had a 10-year portfolio of building projects. It was obvious to all that society's thirst for electricity would keep mounting. The laws that regulated the grid stipulated that the utility keep pace with that thirst. Thus the grid (the nation's and the world's first grid) that had started on Pearl Street with 33 kilowatts was now about to be enhanced by 6 billion watts. The plan called for several new nuclear plants, some to be built up the Hudson River or on an island offshore from New Rochelle, and maybe one on Roosevelt Island in the East River.[26] New York City would have plenty of power.

Allis and Apollo. If we can think of the Apollo mission as an electrical grid attempting to land on the moon, we can also ponder, conversely, the idea of Con Ed's portfolio of new machines as a mission attempting to land in New York City. Apollo might have had to contend with the hostile conditions of space, but Con Ed had to contend with something nearly as hostile—the political conditions in the big city. Rate-increase hearings, fuel purchases, community board meetings, city council elections, labor negotiations, winning building permits, meeting stricter emissions standards—every one of these was a battleground. Landing new dynamos on the surface of New York would not be easy. Even maintaining existing missions, such as Allis, was a daily, combative, draining effort.

What about other grids? In 1969 China's grid is at a standstill as the nation recovers from its cultural revolution. Many universities (even those that decades later would be graduating thousands of engineers) are in 1969 still shut or just reopening. And while Communist China is struggling with the consequences of its red revolution, India is undergoing a green revolution, an effort to develop agriculture in a new way, especially with intense irrigation. Electricity, most believe, is crucial to carrying the scheme forward. Many generators have been installed, per-capita consumption of electricity has improved dramatically, but there are disparities. Some rich farmers get essentially free electricity, while many others get none at all.

In South Vietnam in 1969, amid war the grid is an easy military target. Generators must be defended as if they were forts. Nevertheless, U.S. authorities officially maintain that the war can be won, and David Lilienthal is assigned the task of creating a postwar development plan for the delta region in the south. Lilienthal's proposal for what might have become a Mekong Valley Authority, including hydroelectric facilities, becomes irrelevant as the military situation deteriorates.[27]

In 1969 Amory Lovins is doing research at Oxford. For some time now he has been keenly interested in three topics—the environment, resource use, and international security—and sees them as crucial to future economic development. Where these three topics converge is energy; he wants to write a dissertation about energy. Oxford has a different view. Energy is not a proper academic subject, the university insists, so Lovins resigns his fellowship and heads for greener pastures.

In 1969 the Idaho Power Company is stymied in its effort to build more hydroelectric plants or replace the dam at Swan Falls with a larger facility. It must, however, find new ways to make power. Accordingly, the utility decides solemnly to supplement water power with coal power. It decides to build a huge coal-fired plant in Wyoming.[28]

The moon is also about to receive new electrical equipment. The grid that was being sent there would of course be only a tiny, expensive, provisional version of the grid that cities and nations on Earth have. The delivery vehicle for the lunar grid, *Apollo 11*, roars off into space on July 16, the heaviest thing yet sent into space. The Saturn V rocket, the most powerful engine ever devised, making the loudest man-made sound (if you don't count the sound of a nuclear explosion), is discarded as soon as it finishes its job. For its trip to the moon the composite craft now consists of the command module, where the men are, followed by a tool and supply shed called the service module, and in back a portable garage for the second family vehicle, the lunar module.

Apollo and Allis. The ship going to the moon contains the most advanced electrical equipment money can buy. The command module, containing 2 million parts, is a movable electrical grid zipping away from Earth at an initial 23,000 miles per hour, as fast as humans have ever gone. Apollo's technology cousin, Big Allis, has just as many parts but doesn't travel anywhere. The view from where the astronauts sit in the crew compartment looks like what the dispatcher at Allis's control room would see: a panel covered with indicators, sensor displays, digital readouts, control buttons and knobs, and nearly 600 switches. The available power amounts to about 2,000 watts, about what a large house uses at the dinner hour.

The energy source for this extraterrestrial substation sits behind the astronauts in the service module and consists of fuel cells, devices in which oxygen and hydrogen are combined to make electricity; this same process might someday power hydrogen cars on Earth. No wire runs from Allis to Apollo, but they are still umbilically connected. All Apollo's electrical energy in stored form ultimately comes from Allis or one of its dynamo cousins in a NASA place like Texas or Florida. Apollo might roam far from home, but it was nurtured and fed by the mother grid on Earth.

How much did this earthbound electricity cost in 1969? Depends on where you were and what you were doing. If you were camping and using a flashlight with AA batteries, electricity might cost a hundred dollars per kilowatt-hour. If your power supply were a special-made battery, such as the one in a hearing aid or lunar module, it could cost a thousand dollars or more per kilowatt-hour.

Apollo's battery-powered electricity was expensive. Allis electricity was cheap. Consolidated Edison's newsletter, the one sent around while Buzz Aldrin and Neil Armstrong were at the moon, provides a detailed cost comparison between the price of everyday things in 1946 and 1969. For example, a 1-pound grapefruit went from 6 cents to 14 cents. The cost of an airplane flight from New York to Detroit: up from $22 to $33. Simmons innerspring mattress: up 200 percent. New York subway fare: from 5 cents to 20 cents, a whopping 300 percent increase. The only thing that went down in Con Ed's study (and this is supposed to be a pleasant surprise) was the cost of electricity: *down* 17 percent, to about 4 cents per kilowatt-hour.[29]

In July 1969, as the temporary lords of space hurtle toward the moon, Prince Charles of Britain, humble on bent knee, is invested as Prince of Wales by his mother the Queen at Caernarvon Castle. Watched by millions, the Prince of Wales is soon eclipsed by Senator Edward Kennedy who, accompanied by a young female staffer, accidentally drives off a bridge, sending his car into the water. He struggles to the surface, swims to shore, and dazedly walks away from the scene and does not notify authorities until later the next day. His companion, unfortunately, drowns, and an investigation is launched.

A quarter million miles overhead, Apollo arrives on station and separates into its two component parts. The main craft, *Columbia*, remains parked in an orbit 69 miles above the moon's surface, while the smaller craft, *Eagle*, sinks to an orbit of only 10 miles. *Eagle* is a subcontinent of Apollo but is fully self-contained. Its internal grid, the grid being brought to the moon, is powered by batteries. In the lunar vehicle the two astronauts, Armstrong and Aldrin, are anxious but confident, eager but professional. Both have combat experience, but here they have no rivals, no enemies. No one is shooting at them. An unmanned Soviet probe is lurking in lunar orbit but is not in the way.

Millions might have been following the splendid human thrust into the unknown, but millions more are caught up in the hazards of daily life. At this moment, American soldiers are fighting in Vietnam, El Salvadoran troops are in Honduras, and Israeli fighter jets are over Egypt.

GRID DESCENDING

Allis and Apollo. The giant generator in Queens is the high-water mark of steam-electric technology, just as Apollo represents the peak performance of aeronautical and astronautical engineering. Allis is the most visible asset within the grid founded by Edison, just as Apollo is the prime accomplishment of the NASA system, founded on the work of people like Werner von Braun and Robert Goddard. Both projects—Allis and the combined grids of the world, and Apollo and all the ancillary spacecraft and satellites put into orbit—are human feats on the scale of Egypt's pyramids. On July 20 they are trying to land the pyramid on the moon.

Elevation of 50,000 feet. The landing phase of the mission begins when the descent rocket comes on. Electricity flows to an onboard gimbal, which starts to sense the craft's center of gravity so that the rocket firing can be adjusted to give just the right amount of counterthrust, a process that can be compared to trying to balance an upright broom handle in your palm.[30] Once the gimbal and computer have sorted out this balance business, the rocket can be throttled up. The rocket is being used as a brake, slowing the craft and causing it to slip out of orbit. The rest of the astronauts's ride down will extend 300 miles horizontally, and 50,000 feet vertically and will take 12 minutes. It will be the most important landing since Columbus.

In the vacuum of space there are fortunately no aerodynamic forces straining to rip *Eagle* apart. There will be, however, electrical and guidance problems lying in wait, and the pilots will need every last drop of agility built up over their thousands of hours of training and, in a grander sense, over the million-year evolution of the human brain.

43,000 feet above the moon. Even as it falls, *Eagle* pitches up a bit so that the radar beams shooting out of the craft's lander legs, can better sense the on-rushing lunar plain below. *Eagle* shoots across the landscape in a shallow diagonal, with the bottom of the lander going first. The men just now start to feel their own weight again for the first time in four days. The return of muscle tone is a sensation that no terrestrial grid operator ever experienced.

Allis and Apollo. Both have their own local control rooms—Allis at the Ravenswood plant in Queens and Apollo in the form of its lunar lander. And how about the men inside? Each has his own internal neuroelectrical grid control center, the one between his ears, an organ weighing about 3 pounds. Actually, in the dilute lunar gravity the human brain weighs only half a pound.

In 1969 computers are not that sophisticated. That is not to say they aren't important. Truly at this moment, *Eagle's* onboard computer is vital. It does a lot of the split-second calculations that the overpressed astronauts cannot perform. It keeps track of the approaching surface, initiates appropriate rocket firings, and senses and actuates dozens of other appendage subsystems in an anthill frenzy of electric signals sent and received. All this while staying within a tight budget of watts.

The total situational awareness of the mission at the local level is being dutifully reported to the overseers in Houston and to the pair of humans closer at hand. The onboard computer reveals the spacecraft's status through a variety of alarms and indicator lights. It might not have the aplomb of Hal, the computer running the mission in the movie *2001: A Space Odyssey* (released only the year before), but it does seem at times to have a personality of its own.

Computers and humans. One is made of silicon, the other of carbon. These two elements sit above and below each other in the periodic table and are both gregarious in forming compounds. But carbon-filled human brains—participating in all those symbolic actions like ballroom dancing, exercising grammatical rules, painting marks on cave walls, climbing difficult mountains—have the ability to reinvent, or at least reorient, themselves, whereas silicon computer brains do not alter their behavior except with the help of human-provided software.

35,000 feet. Suddenly things change. Complexity intrudes. Small perturbations have occurred. Let's see if they turn into large disturbances. An alarm goes off and the data screen goes blank. The computer is not keeping up with events. It's been such a smooth ride until now. Why this? Rocket, gimbal, thrusters, radar data are all flooding in and decisions must be made. The craft seems to be proceeding on course, but the astronauts want to know the meaning of the alarm. Houston, what's going on?

The word *telemetry* means metered information traveling far; in this case the data in front of Armstrong and Aldrin also go to Houston, 238,000 miles away. It takes a second and a half to get there, so reality in Texas and reality at the moon are slightly different. The data travel from *Eagle* to an antenna in Goldstone, California, and then over to Houston. Word then goes back in the opposite direction: "Go," meaning proceed. Don't listen to the alarm. Listen to us. The astronauts accept this judgment from afar, but both men look over at the Abort button on their console.

Allis and Apollo. Edwin Nellis, in the Con Edison control room on the night of the 1965 blackout, had only minutes to decipher perplexing electrical readout. Should he stick with the mission—keep the Eastern Interconnection going—or should he abort? His indecision helped snatch the grid away from millions. Now, four years later, Neal Armstrong and Buzz Aldrin, essentially flying a compact, rocket-powered electrical grid dropping down out of the black, must also make sense of perplexing electrical readout. This is no time for Hamlet-like equivocating.

Another warning light comes on, and another jangle flickers through their nervous systems. They loft another query toward Houston. Moments of silence shroud the hurricane of split-second thinking occurring at the far end. Encapsulated in radio waves and sent onward, another "Go" shoots back out of the vacuum. Ignore the computer, the men are being told. The astronauts obey orders. Both have been to war. Aldrin shot down two planes in Korea. Armstrong has himself been shot out of the sky. These men want to get to the moon's surface, but they also want to live. They could activate the Abort button, jettison the lower half of their own craft, and shoot back up toward *Columbia*. But no, they hang tough.

20,000 feet. Slower, lower, closer, more vertical. Engine, thrusters, all seemingly working properly. The horizon now climbs into view in their triangular windows. Another alarm, more tension, and another "Go" from Houston. They must perform a landing in the face of doubtful data.

500 feet. Here's where all that training proves itself. Here's where the bioelectricity flowing through the carbon passengers has to override, or at least supplement, the electrical judgments made in the silicon computer. Software from 1969 has gotten them this far, but no farther. Neil Armstrong appropriates control of the craft. Carbon takes control of the steering wheel away from Silicon. He does not like the view out the window. Damn, but they've overshot the chosen landing field. He slows the descent, stretches out the landing approach, and like a picky homebuyer shopping for something better, flies the thing sideways like a helicopter, away from this rock patch strewn with boulders. He wants a level surface, both for a safe landing and for a safe departure later on.

160 feet. Some of the electrical messages the astronauts receive are ignored, but others have to be taken very seriously. Now the low-fuel light comes on. The extended landing maneuvers are taking their toll. The voice from Houston breaks in to inform them that they have 60 seconds of fuel left. Armstrong is being careful. He wants to set down in just the right place.

30 feet. The silent rocket thrust is kicking up dust now. Houston: 30 seconds of fuel left. Aldrin looks at the Abort button again. As far as we know, nothing living has ever visited this corner of the universe till now. Armstrong eases her down.

The merest mechanical touch is all it takes. The news naturally arrives electrically. A circuit is closed and a signal flows to the dashboard, where the blue contact light comes on. Houston, where the controllers are blue in the face, believes the ship has landed. That's what their monitors tell them, as they sit a quarter million miles away. But they're 1.5 seconds out of touch, and they ardently want confirmation from the guys themselves. After a moment or two by the clock and several heartbeats, the words Houston and the world long for finally come. "Houston, *Tranquility Base* here. The *Eagle* has landed." They weren't a moving vehicle anymore. They were a moon base, a temporary colony, a fixed thing on a worldly body other than Earth.

Among the myriad plans and facts and expectations the astronauts had digested was the prediction that in the instant after touchdown and engine shutdown, the men would feel the sloshing of extra fuel in the tank as the craft settled into repose. But there was no slosh. When the voyageurs finally washed up on the shore of their new land, only 20 seconds of fuel remained in their descent-stage tank. *Eagle* had been running practically on fumes.[31]

TRANQUILITY GRID

In 1969 who has the grid? In China, where they take no notice of the moon landing, some people in cities have it; outside the cities, far fewer. In India it's the same. In India members of parliament spontaneously stand and cheer Armstrong and Aldrin. In Uganda the grid has hardly landed at all. In Ohio the situational awareness of voltage levels might sometimes be faulty, but at least everyone has electricity. As for the moon, the temporary inhabitants there don't have much of a grid—it's not an energy sea but more like an energy puddle nestled in the middle of the Sea of Tranquility—but it's there. And it only has to last another day or so.

The men on the moon have landed but are anything but tranquil. Having rested some hours after their nail-biting descent, Neil Armstrong and Buzz Aldrin are itching to take their walk. The visit won't be official until human footprints have been impressed into the gray carpet outside their door, the debris of 4 billion years of meteor impacts. Armstrong is at the open hatch, trying not to knock things around in the cabin as he exits. He's wearing a special traveling suit, the most expensive garment in history, provisioned with its own electrical network, a home away from home away from home.

Today in Mexico City the grid is converting from a frequency of 50 cycles per second to 60, just like the Yankee grid to the north.

In New York's Central Park and London's Trafalgar Square, crowds watch The Step on giant video screens. One-fifth of all humans alive are watching TV or listening to radio at this moment—a very high degree of tuning in, you might say. The pope has used the Vatican telescope to look at the general lunar landing zone. The emperor of Japan is watching from his private estate. President Nixon is by his telephone, waiting to make what he will call the most historic phone call of all time. Neil Armstrong climbs down his strutted gangplank into the Sea of Tranquility, his boots sinking to the depth of less than an inch into the lunar dust.

At this most watched event in history, electricity plays its part. Consolidated Edison notices that demand for television-consuming power is about 200 megawatts above normal. The surge in Tokyo goes up even more. In Iran, on a rickety black-and-white set, the young Massoud Amin watches the lunar steps. He is so impressed by the spectacle that he decides to become an engineer.

On this momentous occasion, the *New York Times* asks a number of prominent people, all known for their thought-provoking pronouncements, to add some perspective to the already swelling chorus of public sentiment. Most of the essays take the time to face the question of whether the whole exercise has been worth the tremendous expense. For example, Eric Hoffer, who has had an unusual career as longshoreman-turned-philosopher, argues that we would be wrong to spend money only on the necessities of life. "The necessary has never been man's top priority. The passionate pursuit of the nonessential and the extravagant is one of the chief traits of human uniqueness."

The Dalai Lama, asked for a response, cites the Buddhist belief that other civilizations, some of them more advanced than our own, exist in the universe. He hopes that we might someday open communications with these faraway cultures. In the meantime, it would be wonderful if the moon experience led to more mental peace in general.

Putting a man on the moon? Pablo Picasso's attention, as always, is on his pictures. Here is his entire essay (my favorite) on the subject: "It means nothing to me. I have no opinion about it. I don't care."[32]

Flying to the moon is impossible. So says a 115-year-old woman in Japan. The woman is so old that she was alive when a ship came into Edo Bay, a type of armor-plated vessel that had never been seen before. On that occasion the American commander, Admiral Perry, and his men in white suits, demanded that Japan open her port and her markets to commerce with the outside world. When asked for her venerable impression of the Apollo exploit, the old woman says that it couldn't have happened. The moon is a goddess to be worshipped, she says sincerely, not a place to be visited.[33]

But they *have* visited. They are there, at least according to the best electromagnetic telemetry, and the video stream, and the eyewitness testimony of Aldrin and Armstrong. These men on the moon harvest rocks and plant a seismometer on the surface. Then they hoist themselves back into their portable home, their "base," and it's all homeward bound from there. To reduce the weight, they've thrown those heavy packs out the hatch, those back-mounted water-oxygen-electricity supply chests that had sustained them during their hike in the dust. The first electrically recorded scientific result from the moon presents itself. The plop of the discarded backpacks into the lunar dust is picked up by the seismometer.

Apollo and Allis. The part of the electrical grid that remains behind on the Sea of Tranquility, the seismometer, will shortly begin to overheat. Just about that time, back at the electrical grid attached permanently to Queens, New York, Big Allis will develop a "flashover," a sort of short circuit. It will flip off, sending the city into its worst electrical crisis since the 1965 blackout.

It has been said that the Apollo program is the culmination of the Industrial Revolution.[34] It won't be a culmination, however, without a safe trip home. *Eagle* has many redundancies, and why not, since if you're on the moon and a vital component doesn't work it's not as though you can order a replacement. So, for example, the electrical plant for the craft consists of two silver-zinc batteries, either one of which would get the mission done. The one big thing for which there is no redundancy is the ascent rocket that lifts *Eagle* back up for a rendezvous with the mother ship *Columbia*. And the rocket doesn't fire without the enabling electric signal, and the signal doesn't flow forth unless that switch on the circuit breaker is closed, and the astronauts notice now dishearteningly that the switch has broken off, probably when one of the big backpacks swiped against it as they were going out the door for the walk.

In an alternative universe, the switch is not found, the circuit stays open, electricity does not flow, the rocket goes unfired, the astronauts work frantically until their oxygen runs out, and millions of people back on the moon's blue companion world mourn the loss of brave men. In the universe we know, however, things happen differently. Ingenuity, manifested in the astronauts' cerebral electrical grid, wins out. Aldrin and Armstrong use a felt-tip pen in place of the missing switch. The circuit is closed, the countdown proceeds, and the liftoff is perfect.[35] If you had been there, it would have seemed absurd. The rocket's red glare would be evident but no roar. There would be no sound since the flame had no air to set vibrating.

Apollo and Allis. Both represent terminal technology. Look at what happens over the next few years. A few steam-powered electric generators larger than Allis are built, but then the bigger-is-better approach ends. As for Apollo, five more ships would arrive on the moon. A total of six lunar landers are left behind in the dust. A total of 12 men, a corps the size of a jury, stride the gray plains. No one since has paid a call.

Nor should they. It's too expensive under present economic conditions to go back to the moon, much less proceed to Mars. Let Apollo return to Queens, New York, or Uttar Pradesh, India, instead of to the moon. A far more exciting mission than going to the red planet would be going to a green planet. What we need more than Martian footprints is to release human energies for the cause of creating a new type of machine that efficiently delivers electricity without thrusting unwelcome substances into the soil or sky. If we're clever, we ought to be able to have air conditioning (set at moderate temperatures) in Dallas while also keeping the ice sheet attached to Greenland. Let's combine the engineering genius expended on Apollo and Allis, add to it J. P. Morgan's shrewd ability to marshal financial resources, and soften the venture with Thoreau's wise circumspection.

Because of the impact of the United States on fuel markets, global emissions, world trade, and international protocols of all kinds, the American Congress is, whether they want the job or not, an energy planning board with a vast—indeed, planetary—scope for action. Let's hope they recognize this responsibility, take it to heart, and make wise decisions in coming years.

A MEANINGFUL GRID

No, humans don't yet need to return to the moon. The point of going there in the first place has already been made. Ah, but what was that point? What is the meaning of the visit? Decades later we're still trying to find out.

Pablo Picasso didn't care that we had journeyed to the moon and neither did Lewis Mumford. An incessant critic by then of America's involvement in the Vietnam War and in all things militaristic, Mumford insisted, in his sour contribution to the *New York Times*, that the moon landing was no more than a gigantic sporting event. But wasn't that the aim? I would ask him. The moon landing *is* big and it *is* sporting. Why else would those hundreds of millions of transfixed viewers be interested? One-fifth of Homo sapiens were paying attention? Why?

To answer Mr. Mumford I shall quote him his own words. He long held that the main human mission, if we can ascribe a purpose to a species buffeted by blind evolutionary forces, was to become more human:

> But it is not by the light of burning wood that one must seek ancestral man's source of power: the illumination that specifically identifies him came from within. The ant was a more industrious worker than early man, with a more articulate social organization. But no other creature has man's capacity for creating in his own image a symbolic world that both cloudily mirrors and yet transcends his immediate environment. Through his first awareness of himself man began the long process of enlarging the boundaries of the universe and giving to the dumb cosmic show the one attribute it lacked: a knowledge of what for billions of years had been going on.[36]

For billions of years what was happening on Earth was the evolution of living organisms. Toward the end of this long bio adventure, the flourishing of one of those species was helped by ever more elaborately patterned neural firings. These patterns might have led first to an ability to survive glacial nights and tiger attacks and would later have manifested itself as Cro Magnon paintings, Mayan carvings, and the poems of Emily Dickinson. The electrical grid is one of the more recent patterned human contrivances, and it too is both a handy tool and, at some deep level, a complex gesture played out as part of the restless evolutionary game.

When Big Allis landed on New York City or when Apollo landed on the moon, we could see them. But where does human consciousness land? The internal energy sea of the electrical grid can be traced, as can, at least in a physical sense, the neurons of the brain. By contrast, the mind, the great internal electrified mental sea, is harder to locate. With additional applied scientific, artistic, spiritual, and engineering study, more of the answer might come into view. Maybe right now in some corner of India or China a new Faraday is seeing something out of the corner of her eye, experiencing a remarkable flicker of recognition, the kind that leads to a new grid we hadn't thought of before.

NOTES

1 THE GRIDNESS OF THE GRID

1. *New York Post*, August 21, 2003.
2. Homecare Magazine, October 1, 2003.
3. Albert E. Moyer, *Joseph Henry: The Rise of an American Scientist* (Washington, DC, Smithsonian Institution Press, 1997), p. 102.
4. D. K. C. MacDonald, *Faraday, Maxwell, and Kelvin* (Garden City, NJ, Anchor, 1964), p. 51.
5. Henry David Thoreau, *A Week on the Concord and Merrimac Rivers* (New York, Signet Classics, 1961, orig. pub. 1849), pp. 90–101.
6. *Washington Post*, August 17, 2003.
7. *Wall Street Journal*, August 18, 2003.
8. *Technical Analysis of the August 14, 2003 Blackout: What Happened, Why, and What Did We Learn?*, North American Electric Reliability Council, July 13, 2004, p. 1.
9. Dark Sky Association Web site, *www.darksky.org/links/lightpoll.html.*
10. Marufu et al., "The 2003 North American electrical blackout: An accidental experiment in atmospheric chemistry," *Geophysical Research Letters*, vol. 31, p. L13106 (2004).

2 GRID GENESIS

1. Edwin G. Burrows and Mike Wallace, *Gotham* (Oxford, Oxford University Press, 1999), p. 1065.
2. Thomas Hughes, *Networks of Power: Electrification in Western Society, 1880–1930* (Baltimore, Johns Hopkins University Press, 1983), p. 29.
3. Neil Baldwin, *Edison: Inventing the Century* (New York, Hyperion, 1995), p. 106.
4. *New York Times*, December 21, 1880, p. 2; *New York Herald*, December 21, 1880, p. 5.
5. Jill Jonnes, *Empires of Light: Edison, Tesla, Westhinghouse and the Race to Electrify the World* (New York, Random House, 2003), p. 64.
6. Hughes, p. 72.
7. Ibid., p. 42.
8. Baldwin, p. 138.
9. *Edisonia, A Brief History of the Early Edison Electric Lighting System* (New York, Committee on the St. Louis Exposision, 1904).
10. Payson Jones, *Consolidated Edison System Power History* (New York, Consolidated Edison Company, 1940), p. 198.
11. *New York Times*, September 5, 1882, p. 8.
12. Jones, pp 47–49.
13. Ibid., p. 211.
14. *New York Times*, January 3, 1890, p. 8.
15. Jones, p. 220.
16. John J. O'Neill, *Prodigal Genius: The Life of Nikola Tesla* (New York, Washburn, 1944), pp. 43–48; Inez Hunt and Wanetta W. Draper, *Lightning in His Hand: The Life Story of Nikola Tesla* (Denver, Sage Books, 1964), p. 33; Jonnes, p. 93.
17. O'Neill, p. 55.
18. A very handy Web site for explaining common misconceptions about electricity and electrical energy can be found at *http://amasci.com/me.html.*
19. D. K. C. MacDonald, *Faraday, Maxwell, and Kelvin* (Garden City, NJ, Doubleday Anchor Book paperback, 1964), p. 87.

20. Harold I. Sharlin, *The Making of the Electrical Age* (London, Abelard-Schuman, 1963), p. 140.
21. Ibid., p. 145.
22. Nikola Tesla, *Lectures, Patents, Articles*, selected and prepared by Ovjin Popovic, Radoslav Howat, and Nikola Nikolic (Beograd, Yugoslavia, Nikola Tesla Museum, 1965), p. A-104.
23. Tesla, "A New System of Alternating Current Motors and Transformers," p. L1.
24. Marc J. Seifer, *Wizard: The Life and Times of Nikola Tesla, Biography of a Genius* (Secaucus, NJ, Brick Lane Press, 1996), p. 52.
25. Jonnes, p. 123.
26. Henry G. Prout, *A Life of George Westinghouse* (New York, American Society of Mechanical Engineers, 1921), p. 114.
27. O'Neill, p. 79.
28. Harold L. Platt, *The Electric City: Energy and the Growth of the Chicago Area, 1880–1930* (Chicago, University Chicago Press, 1991), p. 246.
29. David E. Nye, *Electrifying America: Social Meanings of a New Technology, 1880–1940* (Cambridge, MIT Press, 1990), p. 37.
30. Reference to the 1893 fair in comparison with a later Chicago fair, *Official Guide: Book of the Fair, 1933* (Chicago, Century of Progress, 1933), p. 27.
31. *Official Guide to the World's Columbian Exposition*, compiled by John Flynn (Chicago, Columbian Guide Company, 1893), p. 56.
32. Ibid., p. 53.
33. Henry Adams, *The Education of Henry Adams* (London, Penguin Classic paperback, 1995, orig. pub., 1918), pp. 325–326.
34. Hughes, p. 134.
35. Ibid., p. 73.

3 MOST ELECTRIFIED CITY

1. Neil Baldwin, *Edison: Inventing the Century* (New York, Hyperion, 1995), p. 178.
2. Samuel Insull, *The Memoirs of Samuel Insull: An Autobiography*, Larry Plachno, ed. (Palo, IL, Transportation Trials, 1992), p. 62.

3. Forest McDonald, *Insull* (Chicago, University of Chicago Press, 1962), p. 54; Insull, p. 63.
4. McDonald, p. 100.
5. Insull, p. 80.
6. David E. Nye, *Consuming Power: A Social History of American Energies* (Cambridge, MIT Press, 1999), p. 18.
7. Nye, p. 18.
8. Samuel Insull, "Centralization of Power Supply," in *Public Utility Economics: A Series of Ten Lectures Delivered Before the West Side Young Men's Christian Association* (New York, YMCA, 1914), p. 96.
9. Harold L. Platt, *The Electric City: Energy and the Growth of the Chicago Area, 1880–1930* (Chicago, University of Chicago Press, 1991), p. 74.
10. David E. Nye, *Electrifying America: Social Meanings of a New Technology, 1880–1940* (Cambridge, MIT Press, 1990), p. 86.
11. Platt, p. 81.
12. Thomas Hughes, *Networks of Power: Electrification in Western Society, 1880–1930* (Baltimore, Johns Hopkins University Press, 1983), p. 204.
13. Platt, p. 153.
14. Gordon Moore quoted in *Crystal Fire: The Birth of the Information Age*, by Michael Riordan and Lilian Hoddeson (New York, Norton, 1997), p. 284.
15. Nye, *Electrifying America*, p. 274.
16. Lewis Mumford, "Bacon: Science as Technology," reprinted in *Interpretations and Forecasts: 1922–1972* (Franklin Center, PA, The Franklin Library, 1981), p. 190.
17. Lewis Mumford, *Technics and Civilization* (New York, Harcourt, Brace World, 1934), p. 16.
18. Ibid., p. 225.
19. Nye, *Consuming Power*, p. 187.
20. Platt, p. 213.
21. Leonard Hyman, *America's Electric Utilities: Past, Present, and Future* (Arlington, VA, Public Utilities Reports, 1994), p. 89.
22. Nye, *Consuming Power*, p. 265.

23. Platt, p. 235.
24. Ibid., p. 218.
25. Thomas Edison, quoted in Nye, p. 242.
26. Nye, p. 272.
27. McDonald, p. 104.
28. Ibid., p. 149.
29. Ibid., p. 195.
30. Insull, *Memoirs*, p. 105.
31. Simon Callow, *Orson Welles: The Road to Xanadu* (New York, Viking, 1995), p. 505.
32. McDonald, p. 275.
33. Platt, p. 279.
34. *New York Times*, October 30, 1929, p. 1.
35. Platt, p. 272.
36. Ibid., p. 272.
37. McDonald, p. 316.
38. Insull, *Memoirs*, p. 94.
39. Samuel Insull, front illustration in *Central-Station Electric Service: Its Commercial Development and Economical Significance as Set Forth in the Public Addresses: 1897–1914*, William Eugene Keity, ed. (Chicago, privately published, 1915).

4 IMPERIAL GRID

1. H. G. Wells, *Russia in the Shadows* (Westport, CT, Hyperion Press, 1973, orig. pub., 1921), pp. 159–160.
2. Jonathan Coopersmith, *The Electrification of Russia, 1880–1926* (Ithaca, Cornell University Press, 1992), p. 72.
3. Vladimir Kartsev, *Krzhizhanovsky*, translation by Oleg Glebov (Moscow, Mir Publishing, 1985), p. 269.
4. A. Markin, *Power Galore: Soviet Power Industry: Past, Present, Future* (Moscow, Progress Publishers, undated), p. 17.
5. Kartsev, p. 271.
6. Coopersmith, p. 155.
7. Markin, p. 24.
8. Leon Trotsky, *My Life: An Attempt at an Autobiography*, introduc-

tion by Joseph Hanson, unlisted translator (New York, Pathfinder Press, 1970 edition), p. 519.

9. Ronald Segal, *Leon Trotsky: A Biography* (New York, Pantheon, 1979), p. 294.
10. Trotsky, pp. 519–520.
11. Harold L. Platt, *The Electric City: Energy in the Growth of the Chicago Area, 1880–1930* (Chicago, University of Chicago Press, 1991), p. 272.
12. William U. Chandler, *The Myth of TVA: Conservation and Development in the Tennessee Valley, 1933–1983* (Washington, DC, Environmental Policy Institute, 1984), p. 12.
13. David E. Lilienthal, *The Journals of David E. Lilienthal*, vol. I (New York, Harper & Row, 1964), p. 18.
14. Walter L. Creese, *TVA's Public Planning: The Vision, the Reality* (Knoxville, University of Tennessee Press, 1990), p. 79.
15. North Callahan, *TVA: Bridge Over Troubled Waters* (London, Thomas Yoseloff, Ltd., 1980), p. 46.
16. *Chicago Tribune*, November 4, 1934, p. 46.
17. Ronald C. Tobey, *Technology as Freedom: The New Deal and the Electrical Modernization of the American Home* (Berkeley, University of California Press, 1996), p. 18.
18. David E. Lilienthal, *TVA: Democracy on the March*, revised edition (New York, Harper, 1953), p. 6.
19. Lilienthal, *TVA*, p. 10.
20. C. Herman Pritchett, *The Tennessee Valley Authority: A Study in Public Administration* (Chapel Hill, University of North Carolina Press, 1043), p. 314.
21. Lilienthal, p. 17.
22. Ibid., p. 17.
23. Ibid., p. 75.
24. Ibid., p. 18.
25. Callahan, p. 59.
26. Chandler, p. 71.
27. Lilienthal, p. 16.
28. Chandler, p. 68.
29. Creese, p. 20.

30. Chandler, p. 9.
31. Thomas P. Hughes, *Networks of Power: Electrification in Western Society, 1880–1930* (Baltimore, Johns Hopkins University Press, 1983), p. 36.
32. Walter Morris, "Consolidated Edison History Update," unpublished memorandum (New York, Con Edison Learning Center, February 13, 1986), p. 3.
33. Ibid., p. 16.
34. Payson Jones, *A Power History of the Consolidated Edison System: 1878–1900* (New York, Consolidated Edison, 1940), p. 203.
35. "The City of Light," brochure at the 1939 world's fair (New York, Consolidated Edision, 1939).
36. David E. Nye, *Consuming Power: A Social History of American Energies* (Cambridge, MIT Press, 1999), p. 187.
37. Ibid., p. 202.
38. Walter Moris, Consolidated Edison History Update, unpublished memorandum (New York, Consolidated Edison, February, 1986), p. 39.
39. "Around the System," internal newsletter (New York, Consolidated Edison, June–July 1965).
40. Interview with electrical engineer Jack Casazza, conducted 1994, IEEE History Center (online at *http://www.ieee.org/organizations/history_center/oral_histories,* p. 13).
41. Morris Dickstein, in *Remembering the Future: The New York World's Fair, 1939 to 1964* (New York, Rizzoli, 1989), p. 26.
42. "Around the System," December 1959.
43. "Around the System," August–September 1965.

5 WORST DAY IN GRID HISTORY

1. *Rochester Democrat and Chronicle,* November 11, 1965.
2. *Toronto Star,* November 10, 1965.
3. *Buffalo Evening News,* November 10, 1965.
4. *Albany Knickerbocker,* November 10, 1965.
5. *Boston Globe,* November 10, 1965.
6. *Providence Journal,* November 10, 1965.
7. *Hartford Courant,* November 10, 1965.

8. *Life* magazine, November 19, 1965.
9. "Around the System," internal newsletter (New York, Consolidated Edison, January 1966).
10. *New York Times*, November 12, 1965.
11. *Scientific American*, May 1978, p. 52.

6 THIRTY MILLION POWERLESS

1. *Variety*, November 10, 1965.
2. *The Night the Lights Went Out*, compilation of blackout news in the *New York Times* (New York, Signet, 1965), p. 42.
3. *The New Yorker*, November 24, 1965.
4. *The Night*, p. 152.
5. Ibid., p. 117.
6. Joseph P. Swidler, *Power in Public Interest: The Memoirs of Joseph P. Swidler* (Knoxville, University of Tennessee Press, 2002), p. 157.
7. "Around the System," internal newsletter (New York, Consolidated Edison, January 1966).
8. Per Bak, *How Nature Works: The Science of Self-Organized Criticality* (New York, Springer-Verlag, 1996), p. 2.
9. B. A. Carerras et al., "Complex Dynamics of Blackouts in Power Transmission Systems," in *Chaos*, September 2004, p. 643.
10. Michael J. Carlowicz and Ramon E. Lopez, *Storms from the Sun: The Emerging Science of Space Weather* (Washington, DC, Joseph Henry Press, 2002), p. 99.
11. Per Bak and Kan Chen, "Self-Organized Criticality," *Scientific American*, January 1991; Massoud Amin, ASME/IEEE/USEA Congressional Staffer Briefing, September 24, 2003.
12. *Report to the President by the Federal Power Commission on the Power Failure in the Northeast United States and the Province of Ontario on November 9–10, 1965* (Washington, DC, FPC, December 6, 1965), p. 9.
13. Lewis Mumford, *The Myth of the Machine, Vol II: The Pentagon of Power* (New York, Harcourt Brace Jovanovich, 1970), p. 236.
14. Lewis Mumford, *The Myth of the Machine, Vol I: Technics and Hu-*

man Development (New York, Harcourt Brace & World, 1967), p. 168.
15. FPC report, p. 33.
16. Thomas J. Bleilock, *Waterside: 1901–2001*, unpublished report (New York, Consolidated Edison, 2001), p. 8.
17. *The Night*, p. 36.
18. Ibid., p. 37.
19. *Life* magazine, November 19, 1965.

7 OVERHAULING THE GRID

1. Alexander Lurkis, *The Power Brink: Con Edison, a Centennial of Electricity* (New York, Icare Press, 1982), p. 65.
2. Allan R. Talbot, *Power Along the Hudson: The Storm King Case and the Birth of Environmentalism* (Toronto and Vancouver, Clark, Irwin, and Co., 1972), p. 4.
3. Lurkis, p. 73.
4. *Fortune* magazine, March 1966.
5. *Wall Street Journal*, August 26, 1968, p. 14.
6. Joseph P. Swidler, *Power in Public Interest: The Memoirs of Joseph P. Swidler* (Knoxville, University of Tennessee Press, 2002), p. 201.
7. Leonard S. Hyman, *America's Electric Utilities: Past, Present, and Future* (Arlington, VA, Public Utilities Reports, 1994), p. 142.
8. Ibid., p. 145.
9. John L. Campbell, *Collapse of an Industry: Nuclear Power and the Condtradictions of U.S. Policy* (Ithaca, Cornell University Press, 1988), p. 4.
10. Hyman, p. 142.
11. Edison Electric Insitute, Electric Perspectives, 77/3 (1977).
12. Amory Lovins, "Energy Efficiency, Taxonomic Overview," in *Encyclopedia of Energy*, vol. 2, (Amsterdam, Elsevier, 2004), p. 384.
13. Discussed by Jonathan G. Koomey, "Debunking an Urban Legend," *The World and I*, Septemer 2002 (online at *www.worldandi.com*).
14. Lorrin Philipson and H. Lee Willis, *Understanding Electric Utilities and De-regulation* (New York, Marcel Dekker, 1999), p. 42.
15. Ibid.

16. Richard Hirsh and Bernard Finn, *Powering a Generation: Power History* #8b (Smithsonian, *http://americanhistory.si.edu/csr/powering*), p. 1.
17. Ibid, p. 2.
18. *The Economist*, July 21, 2001.
19. Amory Lovins, "Electricity Solutions for California," talk presented at Commonwealth Club of San Francisco, July 11, 2001 (Snowmass, Rocky Mountain Institute, 2001), transcript p. 2.
20. *Energy User News*, "Commision Report Faults Independent Generators in California Crisis," February 25, 2002.
21. *The Economist*, August 11, 2003.
22. *New York Times*, February 4, 2005.
23. Bethany McLean and Peter Elund, *The Smartest Guys in the Room: The Amazing Rise and Scandalous Fall of Enron* (New York, Portfolio, 2003), p. 270.
24. Robert Kuttner, *Everything for Sale: The Virtues and Limits of Markets* (New York, Knopf, 1997), p. 11.
25. Ibid., p. 3.
26. Ibid., p. 231.
27. Philipson and Willis, p. 47.

8 ENERGIZING THE GRID

1. Henry Adams, *The Education of Henry Adams* (London, Penguin, 1995, orig. pub., 1918), p. 361.
2. Lorrin Philipson and H. Lee Willis, *Understanding Electric Utilities and De-regulation* (New York, Marcel Dekker, 1999), p. 60.
3. Major assessments of anthropogenic contributions to climate change include: National Research Council, *Surface Temperature Reconstruction for the Past 2000 Years*, June 2006; "Climate Change 2001: The Third Assessment Report of the Intergovernmental Panel on Climate Change," November 7, 2001 (*www.ipcc.ch/press/sp-cop7.htm*); and "Human Impacts on Climate," American Geophysical Union Position Statement, December 2003 (*www.agu.org/sci_soc/policy/positions/climate_changes.html*).
4. *Power Magazine*, July/August 2004.
5. *The Industrial Physicist*, September/October 2004.

6. Amory Lovins, E. Kyle Datta, Thomas Feiler, Karl R. Rabayo, Joel N. Swisher, and Ken Wicker, *Small Is Profitable: The Hidden Economic Benefits of Making Electrical Resources the Right Size* (Snowmass, Rocky Mountain Institute, 2002), p. 43.
7. Ibid., p. 62.
8. Ibid., p. 78.
9. Ibid., p. 141.
10. Philipson and Willis, p. 138.
11. Chris Marnay and Owen C. Bailey, "The CERTS Microgrid and the Future of the Macrogrid," Lawrence Berkeley National Laboratory Report LBNL-55281, August 2004.
12. Amory Lovins, "More Profit with Less Carbon," *Scientific American*, September 2005.
13. Paul Hawken, Amory Lovins, and L. Hunter Lovins, *Natural Capitalism: Creating the Next Industrial Revolution* (Boston, Little, Brown and Co., 1999), p. 35.
14. Ibid., p. 36.
15. "Safe, Secure, Vital," brochure issued by Indian Point Energy Center (*http://www.safesecurevital.org*).
16. "Davis-Besse: The Reactor with a Hole in Its Head," Union of Concerned Scientists fact sheet (*http://www.ucsusa.org/clean_energy/nuclear_safety/*).
17. Richard L. Garwin and Georges Charpak, *Megawatts and Megatons: The Future of Nuclear Power and Nuclear Reactions* (Chicago, University of Chicago Press, 2002), p. 17.
18. "New Health Study: Indian Point Risks Too Great," *Riverkeeper*, Issue 3, Fall 2004 (*http://riverkeeper.org*).
19. Richard L. Garwin, keynote speech at the American Nuclear Society Banquet, November 9, 2002 (*http://fas.org/rlg/02119-ans.htm*).
20. Garwin and Charpak, p. 171.
21. "The Future of Nuclear Energy: An Interdisciplinary MIT Study," chaired by John Deutsch and Ernest J. Moniz (July, 2003).
22. *Physics Today*, December 2003, p. 34.
23. *New York Times*, January 27, 2005.

24. Charles Perrow, *Normal Accidents: Living with High-Risk Technologies* (New York, Basic, 1984), p. 97.
25. Ibid., p. 101.
26. *Nature*, August 10, 2006, p 620.
27. "Moving Forward with Nuclear Power: Issues and Key Factors," Final Report of the Secretary of Energy Advisory Board Nuclear Energy Task Force (Washington, DC, DOE, January 10, 2005).
28. *Physics Today*, May 2005, p. 28.
29. Ari Reeves, with Fredrick Beck, executive editor, "Wind Energy for Electric Power," Renewable Energy Policy Project (July 2003, updated November 2003).
30. Lewis Mumford, "The Opening Future," republished in *Interpretations and Forecasts: 1922-1972* (New York, Harcourt Brace Jovanovich, 1973), p. 431.
31. Garwin and Charpak, p. 35.
32. Ibid.

9 TOUCHING THE GRID

1. Susan M. Stacy, *Legacy of Light: A History of the Idaho Power Company* (Boise, Idaho Power, 1991), p. 9.
2. Henry David Thoreau, *Walden*, with Reader's Guide (New York, Amsco School Publications, 1972, orig. pub., 1854), p. 23.
3. Henry David Thoreau, *A Week Spent on the Concord and Merrimack Rivers* (New York, Signet Classics, 1961, orig. pub., 1849), p. 85.
4. Thoreau, *Week*, p. 115.
5. Thoreau, *Walden*, pp. 73 and 40.
6. Ibid., p. 71.
7. *Idaho Statesman*, June 25, 2004.
8. "Integrated Resources Plan," Idaho Power Company, October 2004, p. 23.
9. Ibid., p. 71.
10. Thoreau, *Week*, p. 311.
11. *Idaho Statesman*, September 23, 2004.
12. Thoreau, *Week*, p. 39.

10 GRID ON THE MOON

1. Davide E. Lilienthal, *The Journals of David E. Lilienthal, Vol VII, Unfinished Business 1968–1981*, edited by Helen M. Lilienthal (New York, Harper & Row, 1983), p. 146.
2. John Brooks, *New Yorker* profile of David Lilienthal reprinted in *Business Adventures* (New York, Wellbright and Talley, 1979), p. 271.
3. Lilienthal, p. 159.
4. Amory Lovins, "Energy Efficiency, Taxonomic Overview," in *Encyclopedia of Energy*, vol. 2, (Amsterdam, Elsevier, 2004), p. 389.
5. "Electricity Technology Roadmap: Meeting the Central Challenges of the 21st Century: 2003 Summary and Synthesis," Electric Power Research Institute (Palo Alto, EPRI, 2003), p. 5.
6. "Power in Uganda: A Study of the Economic Growth Prospects for Uganda with Special Reference to the Potential Demand for Electricity," Uganda Electricity Board (London, The Economist Intelligence Unit, 1957), p. 121.
7. Great Lakes Region: Burundi, Kenya, Rwanda, Tanzania, and Uganda, Country Analysis Briefs, Energy Information Agency (Washington, DC, DOE, 2004).
8. "Technical Analysis of the August 14, 2003 Blackout: What Happened, Why, and What Did We Learn?," North American Electric Reliability Council (Princeton, NERC, 2004), p. 1.
9. *New York Times*, August 16, 2003.
10. Ibid., September 29, 2003.
11. Electricity Technology Roadmap, pp. 1–6.
12. Pestonji D. Mahaluxmivala, *History of the Bombay Electric Supply & Tramway Company* (Bombay, BEST, 1936).
13. Sachchidanand, *Electricity and Social Change* (New Dehli, Janaki Prakashan, 1983), p. 109.
14. *Gulf News*, May 27, 1999.
15. India Country Analysis Brief, Energy Information Agency (Washington, DC, DOE, 2004).
16. Report on the Grid Disturbance in Northern Region (2.1.2001), Indian Ministry of Power (*http://www.powermin.nic.in/report/nrd_opt1.htm*).

17. Xinhua General Overseas New Service, June 26, 1990.
18. *New York Times*, December 16, 2002.
19. National Energy Grid: India, Global Energy Network Institute, June 18, 2003.
20. Jeff Goodell, *Big Coal: The Dirty Secret Behind America's Energy Future* (Boston, Hougton Mifflin, 2006), p. xii.
21. Vaclav Smil, *China's Past, China's Future: Energy, Food, Environment* (New York, Routledge Curzon, 2004), p. 24.
22. Jared Diamond, *Collapse: How Societies Choose to Fall or Succeed* (New York, Penguin, 2005), p. 364.
23. Great Lakes Region: Burundi, Kenya, Rwanda, Tanzania, and Uganda, Country Analysis Briefs, Energy Information Agency (Washington, DC, DOE, 2004).
24. Lewis Mumford, *The Myth of the Machine, Vol I: Technics and Human Development* (New York, Harcourt Brace & World, 1967), pp. 33 and 35.
25. Buzz Aldrin and Malcolm McConnell, *Men from Earth* (New York, Bantam Books, 1989), p. 163.
26. "Around the System," internal newsletter (New York, Consolidated Edison, March 1969).
27. David E. Lilienthal, *Foreign Affairs*, January 1969.
28. Susan M. Stacy, *Legacy of Light: A History of the Idaho Power Company* (Boise, Idaho Power, 1991), p. 175.
29. Richard P. Hallion and Tom D. Crouch, eds., *Apollo: Ten Years Since Tranquility Base* (Washington, DC, Smithsonian Institution, 1979), p. 128.
30. *New York Times*, July 21, 1969.
31. Aldrin and McConnell, p. 238.
32. *New York Times*, July 21, 1969.
33. *Japan Times*, July 22, 1969.
34. Aldrin and McConnell, p. 232.
35. Ibid., p. 244.
36. Mumford, p. 30.

ACKNOWLEDGMENTS

This book would not exist if it weren't for my agents Gail Ross and Howard Yoon and my editor Jeffrey Robbins at the Joseph Henry Press, so I thank them first.

I am particularly grateful to those who read parts or the whole of this book in draft form and made helpful comments: Neal Singer, William Sweet, Keay Davidson, Martha Heil, Sidney Perkowitz, Lawrence Papay, John Ahearne, Mark Schewe, and Roy Schewe.

For multiple lengthy conversations on a variety of topics relating to the world of electrical energy, I pay respectful thanks to three conspicuously knowledgeable men: Amory Lovins, chief executive officer of the Rocky Mountain Institute; Massoud Amin, director of the Center for the Development of Technological Leadership at the University of Minnesota; and Paul Grant of the Electric Power Research Institute.

I appreciate the efforts of those at the Joseph Henry Press who helped to turn the text into a book: Lara Andersen, Sally Stanfield, Barbara Bodling O'Hare, Estelle Miller, Charles Baum, Matthew Litts, Ann Merchant, and Jessica Henig.

For making possible my visits to several sites described in the book, for answering numerous questions, and for their warm hospitality, the following individuals get my thanks: Ed Yutkowitz and Erwin Schaub at KeySpan Corporation; Len Middleton at Consolidated Edison; Kathleen McMullin at Entergy Corporation; David Kalson at Ruder Finn; Mark Schewe, Dennis Lopez, Gary Felton, James Terrell, James Miller, and Marsha Leese at Idaho Power Company; and Gene Fadness at the Idaho Public Utility Commission.

An author can't know enough about his subject. I benefited from enlightening conversations with experts on numerous technical matters. For this I thank Walter Baer of the RAND Corporation, Ben Carerras of Oak Ridge National Laboratory, Bruce Wollenberg of the University of Minnesota, Richard Hirsh of Virginia Polytechnic Institute, Mike Tidwell of the Chesapeake Action Climate Network, Edwin Lawless at the Potomac Electric Power Company, Bill LeBlanc, Linda Mundy, and Anthony Fainberg.

Writers often save their last and best appreciation for their family members, who, positioned in the closest possible orbital arrangement with the author and his book, exert the greatest personal benevolent influence. This is true for me too, and I thank them heartily for their proximity, patience, and many kindnesses.

INDEX

C

P

R

U

V

W

Y

Z